# Plastics Product Design

**Scrivener Publishing**
100 Cummings Center, Suite 541J
Beverly, MA 01915-6106

*Publishers at Scrivener*
Martin Scrivener (martin@scrivenerpublishing.com)
Phillip Carmical (pcarmical@scrivenerpublishing.com)

# Plastics Product Design

by

Paul F. Mastro

Scrivener Publishing

WILEY

Co-published by John Wiley & Sons, Inc. Hoboken, New Jersey, and Scrivener Publishing LLC, Salem, Massachusetts.
Published simultaneously in Canada.



For general information on our other products and services or for technical support, please contact our Customer Care Department within the United States at (800) 762-2974, outside the United States at (317) 572-3993 or fax (317) 572-4002.

Wiley also publishes its books in a variety of electronic formats. Some content that appears in print may not be available in electronic formats. For more information about Wiley products, visit our web site at www.wiley.com.

For more information about Scrivener products please visit www.scrivenerpublishing.com.

Cover design by Russell Richardson

***Library of Congress Cataloging-in-Publication Data:***

ISBN 978-1-118-84271-3

Printed in Singapore

MWEP407971241025

# Contents

Preface xi

PART 1: Plastics as a Design Material 1

1 Introduction to Plastics Materials 3
1.1 History of Plastics 3
1.2 Definition of Plastics 5
1.3 Thermoplastics and Thermosets 5
1.4 How Plastics are Made 6
1.5 General Plastics Properties 7
1.6 Plastics Feedstocks and Volumes 8

2 Properties of Plastics 11
2.1 Molecular Weight and Molecular Weight Distribution 13
2.2 Melt Flow Index 16
2.3 Molecular Structure of Polymers 16
2.4 Thermal Properties of Plastics 17
2.5 Physical Properties of Plastics 24
2.6 Electrical Properties 28
2.7 Flammability 29

3 Overview of Plastics Materials 31
3.1 Polyethylene 32
3.2 Polypropylene 35
3.3 Polystyrene 37
3.4 Polyvinyl Chloride 39
3.5 Engineering Plastics 41
3.5.1 Cellulosics 41
3.5.2 Polymethyl Methacrylate (Acrylic) 42
3.5.3 Polycarbonates 43
3.5.4 Polyamides (Nylon) 45
3.5.5 Polyoxymethylene (Acetal) 46

3.5.6 Thermoplastic Polyesters 47
3.5.7 Fluoropolymers 48
3.5.8 High Performance Polymers 49
3.5.8.1 Polyphenylenes 50
3.5.8.2 Polysulfones 50
3.5.8.3 Polyaramids 51
3.5.8.4 Polyarylether Ketones 51
3.5.8.5 Liquid Crystal Polymers (LCPs) 52
3.5.8.6 Thermoplastic Polyimides 53
3.5.8.7 Polybenzimidazole 53
3.6 Thermoplastic Elastomers 54
3.7 Biopolymers 55
3.7.1 Polylactic Acid 55
3.7.2 Polyhydroxyalkanoates 56
3.7.3 Polybutylene Succinate 56
3.8 Thermosets 56
3.8.1 Phenolics 57
3.8.2 Amino Plastics 57
3.8.3 Epoxies 59
3.8.4 Thermoset Polyesters 60
3.8.5 Thermoset Polyurethanes 61
3.8.6 Polydicyclopentadiene 62
3.8.7 Thermoset Polyimides 62
3.9 Fillers and Reinforcements 62

**4 Process Overviews, Advantages and Constraints 65**
4.1 Extrusion 66
4.2 Injection Molding 69
4.3 Extrusion Blow Molding 76
4.4 Injection Blow Molding and Stretch Blow Molding 78
4.5 Compression Molding 81
4.6 Transfer Molding 82
4.7 Rotational Molding 82
4.8 Reaction Injection Molding 85
4.9 Thermoforming 85
4.10 Filament Winding 87
4.11 Pultrusion 89
4.12 Additive Manufacturing (3D Printing) 90
4.13 Other Prototyping Processes 92

**5 General Design Considerations** **93**
5.1 Shrinkage 93
5.2 Dimensional Tolerances 94
5.3 Draft 98
5.4 Gating 100
5.5 Coring and Holes 102
5.6 Rib Design 106
5.7 Color and Appearance 107
5.8 Chemical Resistance 109
5.9 Weathering and Environmental Effects 111
5.10 Recycling and Recycling Codes 112

**PART 2: Plastics Product Design** **115**

**6 Structural Components** **117**
6.1 Rigidity and Strength 118
6.2 Creep 120
6.3 Fatigue 130
6.4 Torsion 131
6.5 Impact 134
6.6 Other Elevated Temperature Considerations 137

**7 Enclosures** **139**
7.1 Cosmetics 140
7.2 Structural Support 142
7.3 Ventilation 148
7.4 Flammability 149
7.5 Electrical Considerations 152

**8 Packaging and Containers** **157**
8.1 Impact and Tear Resistance 157
8.2 Strength and Rigidity 158
8.3 Barrier Properties 158
8.4 Packaging Processes 162
8.5 Printing and Decorating 165

**9 Snap Fits and Hinges** **169**
9.1 Snap Fit Designs 170
9.2 Design of Cantilever Snaps Using Classical Beam Theory 172

9.3 Assembly and Disassembly 180
9.4 Non-Rectangular Cantilevered Beams 186
9.5 Effects of Stress Concentration 186
9.6 Annular Snap Fits 187
9.7 Manufacturability 190
9.8 Plastic Hinges 192

**10 Plastic Gears** **195**
10.1 How Gears Work 196
10.2 Types of Gears 198
10.3 Terminology 201
10.4 Gear Tooth Loading 203
10.5 Contact Stress 208
10.6 Gear Tolerances 209
10.7 Gear Tooth Design 211
10.8 Gear Mesh Conditions and Operating Distances 213
10.9 Software 216
10.10 Prototyping 217
10.11 Gear Manufacturability 217
10.12 Gear Materials 221

**11 Bearings** **223**
11.1 Wear 225
11.2 Bearing Life and Performance 228
11.3 Bearing Design 230
11.4 Bearing Materials 230

**12 Pressure Vessels and Pipes** **233**
12.1 Pipe 234
12.2 Miner's Rule 237
12.3 Other Pressure Vessels 239
12.4 Other Types of Pressure Vessels 243
12.5 Material and Manufacturing Considerations 243

**13 Plastic Optics** **247**
13.1 Optical Fundamentals 247
13.2 Mirrors 252
13.3 Light Pipes 254
13.4 Lenses 254
13.5 Manufacturing Processes for Optical Components 256
13.6 Measuring Techniques 257

**14 Joining Techniques** **259**
14.1 Threads and Threading 260
14.2 Self-Tapping Screws 263
14.3 Metal Inserts 265
14.4 Ultrasonic Welding 268
14.5 Vibration and Hot Plate Welding 272
14.6 Spin Welding 274
14.7 Solvent and Adhesive Bonding 275
14.8 Bolt and Screw Assembly 278

**15 Product Design Process** **281**
15.1 Design Process 281
15.2 Material Selection 289
15.3 Design Services 289

**Appendix A**
Thermal Properties of Selected Generic Materials 293

**Appendix B**
Properties of Selected Structural Components 295

**Appendix C**
Common Abbreviations for Plastic Materials 297

**References** **299**

**Index** **303**

# Preface

My first job as a plastics engineer fresh out of college was working for a large custom molder. That gave me a wonderful opportunity to be involved in the design and manufacture of a wide array of different products spanning a number of different industries. What that experience taught me was that in the myriad of different products that I got to work on, the basic engineering principles were the same, but each different industry or product group had constraints and requirements unique to them and often used very specialized engineering techniques that one would not necessarily be aware of if they did have experience in that industry. Working on different products every day gave me an appreciation for how important this knowledge was and how difficult it could be to quickly acquire it. This was reinforced throughout my career as I worked in engineering and product design in a number of different industries. The main purpose of this book is to convey the knowledge I obtained in all of these experiences to help facilitate the design process for people involved in designing or manufacturing similar kinds of parts.

Plastic materials provide the design engineer a wide variety and ever increasing number of possible solutions to some of the most difficult design problems for an ever widening scope of applications. The purpose of this book is to provide the reader with a basic understanding of the range of plastics materials, properties, and processes available to them and an understanding of how to design a variety of basic components from plastic materials.

While the basic design principles that will be discussed in the first section of the book are applicable to the design of any plastic part, each product type or industry has specific or unique requirements, many of which will be discussed in detail in the chapters in part two that should help someone unfamiliar with them to begin to develop a part design.

This book is intended for use by designers who have had limited or no experience with plastics materials as well as a more experienced designer who is designing a part for a use, process or an application that they are not familiar with. Also, the book has an extensive discussion of materials and processes that will provide a solid introduction to plastics for anyone.

In Section 1, Plastics as a Design Material, the reader is provided with an introduction to plastics as a design material. The section is introduced with a brief history of the plastics industry and a discussion of general plastics material properties. The next two chapters will give an overview of the plastics materials commonly in use today and a discussion of a variety of processes available to the designer to make a part along with the design considerations each process will entail. This section also includes a discussion of useful prototyping processes, including advantages and disadvantages of each. Finally, chapter 5 will discuss general design considerations that are applicable to most plastics product designs.

In Section 2, Plastics Product Design, the discussion will turn to the specific design and manufacturing requirements for a number of different product types and components. This section starts with an introduction to plastic materials being used as structural components, where many basic mechanical engineering principles are reviewed as well as a discussion of how they need to be adapted to account for the viscoeleastic behaviour of plastics materials. Chapters follow on enclosures, packaging, gears, hinges, bearings, snap fits, pressure vessels, including pipe, and optical components. These sections will discuss the general considerations that are relevant to these applications as well as specific insights about each particular application. Discussions of plastic joining techniques that are applicable across all product groups will layout the design choices available for putting these components together. The book concludes with a discussion of the product development process and role of the past design and designer in this process.

This book is the result of my experiences as a design and manufacturing engineer and is intended to provide the reader with a basic understanding of plastics materials and processes and to provide a resource to assist them with the design of a number of different components.

I would like to thank my wife Lois for her help and support throughout my career and in this endeavour.

# PART 1

# PLASTICS AS A DESIGN MATERIAL

Plastics materials have a number of unique properties that allow a wide variety of solutions to many design problems. The nature of some of these properties requires the designer to approach the application of these materials to a product design a little differently than many traditional design materials. This section will review common materials and processes and look at some of the general engineering approaches that need to be taken in developing a plastic product design.

# 1

# Introduction to Plastics Materials

In this chapter we will briefly discuss the history of plastics, examine what plastics are, how they are made and some of the general properties of plastics materials. We will also look at the overall size of the plastics industry today.

## 1.1 History of Plastics

It is hard to imagine a world without plastics, but plastics are a family of relatively new materials and have been around for a little more than 100 years. The start of the plastics industry dates back to 1868 when John Wesley Hyatt, in search of an alternate material to ivory for billiard balls, discovered celluloid, the first commercially successful plastic material. Celluloid also found application in photographic still and movie film and shirt collars and buttons. It is still in use today to make ping pong balls.

Celluloid was a modified naturally occurring polymer, cellulose. In 1907 Dr. Leo Baekeland, through a condensation reaction of phenol and formaldehyde, invented phenolic, the first plastic produced entirely from synthetic materials. This was an easily moldable, cost effective material that became widely used in electrical components and general moldings. Its major limitation was that it was only available in dark colors. This problem was solved in 1929 when American Cyanamid Company introduced urea formaldehyde thermoset molding compounds which could be produced in a wide array of colors.

In 1934 Dr. Wallace Carothers, working for DuPont, invented nylon. This is notable because he was hired to develop a synthetic material to replace silk and he developed a polymer to meet this specific need, a first for polymer chemists.

The first inorganic polymer, polytetrafluoroethylene, more commonly known as Teflon®, was discovered by another DuPont chemist, Dr. Roy Plunkett, in 1938.

Throughout the 1940s thermoset materials dominated the plastics market, but starting in the 1950s new thermoplastic materials and processes began to take over. The first commercial reciprocating screw injection molding machine appeared in Germany in the mid-1950s from Ankerwerk. Due to its ability to produce significantly improved thermoplastic melts, numerous manufacturers around the world soon offered their own versions. Injection-molded thermoplastics started to replace many thermoset applications and many new opportunities for growth were found.

In 1953, the first reinforced plastic car bodies appeared in the Chevrolet Corvette [1], and plastics continue to make inroads in the auto industry as their low costs and high strength-to-weight ratios help engineers meet ever-increasing fuel economy requirements. Use of plastics materials to reduce the weight of cars is a major strategy of the automobile industry as they strive to meet the US National Highway Traffic Safety Administration 2025 CAFE (Corporate Average Fuel Economy) standards of 54.5 miles per gallon by 2025.

Advances in polymer chemistry and catalysts now allow polymer chemists to scientifically develop plastic materials with specific properties to meet the needs of specific applications. Stereospecific catalysts like Zeigler Natta catalysts and metallocene catalysts can help control how and where the molecules attach to one another. Ruthenium catalysts enable ring opening metathesis polymerization, which has opened the possibilities of new families of high performance polymers.

This is allowing plastics to move into areas of much more demanding functional requirements and to be used in a wide array of engineering applications. Plastic materials are not only used in housewares, toys and packaging, but also aerospace, construction, electronics, transportation and industrial applications.

## 1.2 Definition of Plastics

What actually is a plastic material? There are many similar definitions used, but for our purposes, plastics are materials that are composed of large molecules that are synthetically made and, under the proper conditions, can be readily formed or molded into the desired shape. The large plastic molecules are called polymers from the Greek words poly, which means many, and meros which means units. The polymer is made up of many smaller molecules called monomers which are joined together through chemical bonding, generally through either a condensation or an addition polymerization reaction. The chemical properties of the monomer will determine if and how it can form into a polymer, as well as what properties the finished polymer might have.

## 1.3 Thermoplastics and Thermosets

Plastics are divided into two basic families, thermoplastics and thermosets. Thermoplastics are materials that when heated will soften and flow, allowing the polymer chains to slide over one another, and when cooled, they will harden. This process can be repeated many times. This allows thermoplastic materials to be easily recycled and reused. A thermoset material will soften and flow when it is heated, but additional heat will cause a chemical reaction called crosslinking to occur. In crosslinking, chemical bonds form between the polymer chains. This crosslinking reaction locks the polymer chains together and prevents them from sliding over one another, causing the polymer to harden. This process is irreversible. As a result, parts made from thermosets cannot be easily recycled.

Figure 1.1 shows the differences in how thermoplastics and thermosets respond to changes in temperature. At lower temperatures (upper left on the chart) both thermoplastics and thermosets are solids (usually – although a thermoset resin can start out as a liquid). As the temperature is increased, the viscosity (resistance to flow) of both materials will lower until they go from a solid state to a viscous (thick)

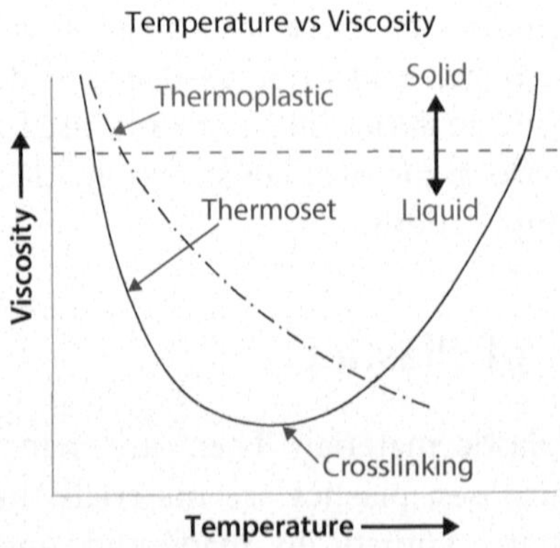


**Figure 1.1** Temperature vs Viscosity.

liquid. As the temperatures continue to increase, the viscosity of both materials continues to decrease. With thermoplastics, this drop in viscosity continues as temperatures increase (up until the chemical bonds start to disassociate and the polymer begins to degrade). With thermosets, the material will rapidly drop in viscosity with increasing temperature until chemical crosslinks start to form between the chains. At that point, viscosity will increase as the crosslinks restrict the polymer chains from sliding past one another. With thermoplastics, we can drop the temperature and reverse the process. With thermosets, once the crosslinks start to form, we cannot reverse the process and the material is said to be cured.

## 1.4 How Plastics are Made

Polymers are generally formed by two different mechanisms, addition polymerization and condensation polymerization. In addition polymerization the monomer is introduced into the reactor and a chemical initiator is added to create a free radical. The free radical will bond with another monomer, creating a new free radical at the same time. This process will repeat, causing the polymer chain to continue to grow. The growth process will continue until the monomer supply is exhausted, two free radicals meet, or a quenching molecule(s) is added to stop the reaction. Addition polymerization processes can be scaled up to a very large size, has no byproducts (the initiator is consumed by the reaction) and are very economical. Addition polymerization is also called chain-growth polymerization. The common high volume commodity plastics (see Chapter 3) are manufactured by this method. In condensation polymerization two different monomers are introduced into the reactor vessel. One end of molecule 1 will react with one end of molecule 2 and usually as part of this process a water molecule is also created (hence the name "condensation polymerization"). Newly formed molecule 1–2 reacts with either a molecule 1 or a molecule 2, extending the polymer chain. The reaction is ended by either adding a quenching molecule or cooling the process down to stop the reaction. Usually, catalysts are added to help drive and control the reaction. Unlike initiators, catalysts are not consumed by the reaction and must be removed from the finished polymer, along with the water that was produced in the reaction. Many of the engineering polymers are made by condensation polymerization. This is generally a more expensive process and will increase the cost of plastics made by this method. Condensation polymerization is also called stepwise polymerization.

## 1.5 General Plastics Properties

There are a large number of commercial polymers available, offering a wide range of properties that can be used in a very diverse set of applications. In general, plastics offer a low density resulting in a low specific weight (weight per unit volume) giving a high strength to weight ratio over competing materials. This characteristic is finding value for weight reduction in transportation applications, such as automobiles, aircraft and high-speed trains, to improve fuel economy. Plastics typically have low melting points which help processing but limit applications in high temperature environments. Although plastics do have lower melting temperatures than many competing materials, there are plastics that can withstand some surprisingly high temperatures, allowing them to be used in under the hood automobile applications and for sterilizable parts. They have a low thermal conductivity which makes them ideal for many thermal insulation applications. They also have good electrical insulative characteristics which make them key materials in a wide range of electrical applications, including wire and cable insulation, tool housings, circuit boards and components and household appliances. Plastics are available with optical clarity and can be made in an infinite array of colors. Lastly, one of the most significant general properties of plastics is that of relatively low cost, which has led plastics to be used in a number of high volume, low cost applications such as packaging materials, containers and insulation. Although these low cost polymers can be obtained for less than a dollar per pound, many higher performance polymers will cost from more than $2.00 per pound up to over $100 per pound.

## 1.6 Plastics Feedstocks and Volumes

The raw material feedstock for today's commercial plastics come principally from petroleum (currently about 2% of petroleum production goes into the manufacture of plastic materials) and natural gas, although there is considerable movement into renewable feedstock materials such as soybeans, castor oil beans, corn, wood fibers and algae. The cost of the polymerized plastic is driven by the costs of the basic feedstocks, the cost of operating the polymerization process, and the volume of the material being produced. Plastics production is a globalized industry and this leads to local feedstock economics influencing how a specific polymer might be made. An example of this would be one of the highest volume plastics produced in the world, polyvinyl chloride (PVC). PVC is made by the addition polymerization of vinyl chloride monomer. In the Middle East and North America, because of the abundance of cheap natural gas, the first step in the process is to make ethylene from natural gas, which is then used to make ethylene dichloride and then VCM. In Europe, where natural gas is more expensive, the ethylene is made from petroleum. In China, where neither petroleum nor natural gas is cheap, they use the carbide process, where cheap coal is the feedstock that leads to the VCM.

We extend the range of properties available in plastics through a compounding process where we take the basic polymer and mix in a variety of additives to enhance and extend properties. We can add ultraviolet absorbers to improve performance in outdoor exposures, flame retardants to improve flammability, thermal stabilizers to improve heat stability and a variety if colorants, including pearlescents and fluorescents, to give a tremendous array of cosmetic effects. By adding in one or more of a wide variety of available reinforcements we can create a class of plastics called composites, which can yield substantial property enhancements, making plastics one of the primary materials used in advanced aircraft such as the Boeing Dreamliner and the BMW electric car.

Plastic volumes worldwide are approximately 650 billion pounds a year. The largest user is China followed by Europe and North America. The United States uses over 100 billion lbs of plastics per year. The single largest used plastics material is polyethylene. Worldwide consumption is over 180 billion pounds in many diverse application from packaging films and bags to milk bottles to gas and water service pipe. The annual volume of plastic surpassed that of steel in 1979 and continues to grow.

# 2

# Properties of Plastics

Each plastic material exhibits a unique array of properties. These properties are determined by the basic chemical makeup and structure of the monomer, the size and variations in size of the polymer chains that make up the polymer and the interactions of one polymer chain with another.

It is assumed the reader has a basic understanding of chemistry, but to facilitate our following discussion of plastic materials, let's first review some polymer chemistry basics. As was discussed in Chapter 1, polymers or plastics are made up of long chains of molecules chemically bonded together. The smallest building block molecule for any polymer is called the monomer. We can represent a polymer with the general formula $-(\mathbf{X})-_{\mathbf{n}}$, where **X** is the molecule or building block of the polymer (the monomer) and **n** represents the average number of monomer units in a polymer chain. The bar "–" represents a pair of electrons which bond the two adjacent molecules together. Most plastics are primarily bonded together by covalent chemical bonds. This means that the adjacent atoms share electrons to fill their orbitals.

A secondary type of bonding called polar bonding (also called hydrogen bonding or dipole bonding) has considerable influence over a number of key properties of a plastic material, including chemical resistance and permeability. Covalent bonding is the sharing of electrons between atoms, but because some atoms are more electronegative than others (for example, oxygen is more electronegative than carbon), the shared electrons will spend more time around the more electronegative atoms, giving them a slightly negative charge.

Figure 2.1 shows part of an ethylene vinyl acetate molecule (EVA). Because the two shared electrons pairs, shown by the two bars between the carbon (C) and the oxygen (O), will be slightly more attracted to the more electronegative oxygen, that end of the molecule has a slightly negative charge. We show this with the notation $\delta^-$. This will affect how the molecule behaves when it comes into proximity with other molecules like itself and also when it comes into proximity with other materials. When this material is made into a film, the tendency of the negative and positive ends to attract, cause the film to stick to itself or "cling."

The main polymer chain is called its backbone and typically is made up of carbon atoms but can also have other atoms such as nitrogen or silicon or functional groups such as benzene rings. If the backbone chain consists only of single atoms bonded together in a chain, that polymer is said to be aliphatic. If the backbone also contains some benzene rings attached to it, that polymer is said to be aromatic. A benzene ring is a hexagonal structure of 6 carbon atoms and 6 hydrogen atoms with alternating single and double bonds between the carbons. Note that a benzene ring could be attached to the side of a chain and not directly in the chain. In that case we still would consider the polymer to be aliphatic.

Looking at the general formula for polymers discussed above, $-(X)-_n$, where **n** is equal to the average number of repeating monomer units, **n** can be a very large number, in the thousands of units. The length of the polymer chain is determined by **n**. A simple measure that is

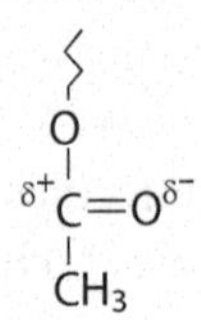


**Figure 2.1** Molecular polarity.

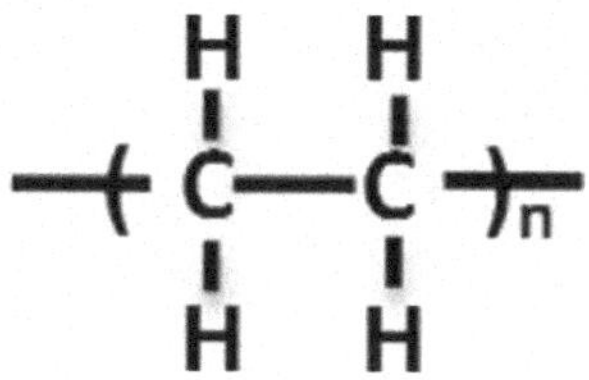

**Figure 2.2** Polyethylene.

commonly used for **n** is the molecular weight. The molecular weight of a polymer is the sum of the atomic weights of each of the atoms in the molecule. For example, for the simplest polymer, polyethylene (shown in figure 2.2), we can calculate the molecular weight by going to the periodic table and find that the hydrogen atoms have an atomic weight of 1 (grams/mole) and the carbon atoms a weight of 12, so the atomic weight of the monomer would be 4 hydrogen atoms × 1 plus 2 carbon atoms × 12 for a molecular weight of 28 for the monomer (the unit of measure is grams/mole but common practice is to just show the molecular weight as just the number, with the measurement units implied). The molecular weight of the polymer would be the monomer weight of 28 multiplied by **n**, the number of repeating units in the polymer.

## 2.1 Molecular Weight and Molecular Weight Distribution

Any real sample of a plastic material will be made up of many polymer chains of varying lengths. The molecular weights reported for any given polymer is actually a statistically generated average that we call the average molecular weight ($\mathbf{MW_{ave}}$). The range of molecular weights in a given sample is called the molecular weight distribution (**MWD**). If we plotted this distribution, it could be normal, skewed or bimodal. Figure 2.3 gives a graphical example of MWDs.

Two polymer samples, like those in Figure 2.3, can have different MWDs but the same $\mathbf{MW_{ave}}$.

The $\mathbf{MW_{ave}}$ can be calculated in a number of different ways but the most common way is by what is known as the number average molecular weight $\mathbf{M_n}$. $\mathbf{M_n}$ can be calculated by using the Equation 2.1

$$M_n = \Sigma \frac{N_i M_i}{\Sigma N_i} \tag{2.1}$$

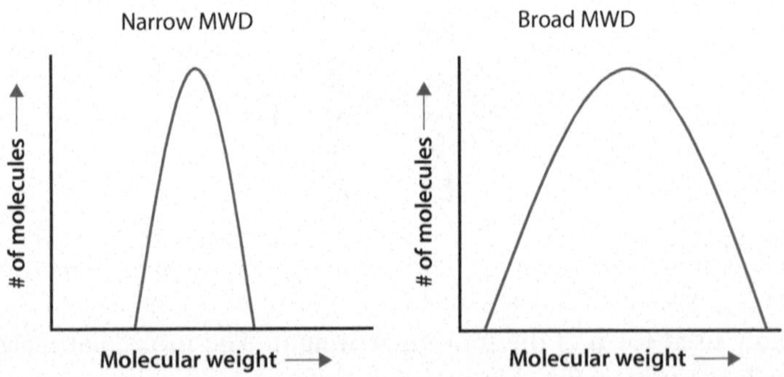


**Figure 2.3** Molecular weight distribution.

where $\mathbf{M_i}$ is the molecular weight of a particular molecule in the sample and $\mathbf{N_i}$ is the number of molecules in the sample with that molecular weight. To illustrate, let's use a simple example. We have taken a small sample of polymer and found the following molecules in it:

| Number of molecules | Molecular weight |
|---|---|
| 1 | 10000 |
| 3 | 25000 |
| 5 | 50000 |
| 4 | 75000 |
| 1 | 100000 |

Substituting in Equation 2.1 we find $\mathbf{M_n}$ = 52,500. Typical molecular weights of commercial polymers range from 30,000 to over 1,000,000 grams/mole.

Other methods of calculating $\mathbf{MW_{ave}}$ include a weight average molecular weight, $\mathbf{M_w}$, which averages the weights of the individual molecules which relate better to some properties such as permeation, and the viscosity average molecular weight $\mathbf{M_z}$.

$\mathbf{MW_{ave}}$ and **MWD** can be determined experimentally through a variety of experimental techniques, including gel permeation chromatography and end-cap analysis. From this, the average number of monomer units in the polymer, **n**,can be determined by dividing the experimentally determined MW by the calculated molecular weight of the monomer unit. The average value of **n** for a polymer is sometimes referred to as the degree of polymerization.

In our example above, the calculated number average molecular weight $\mathbf{M_n}$ is 52,500 but the polymer sample contains some chains much shorter and some much longer. This **MWD**, as well as the absolute $\mathbf{MW_{ave}}$ will affect polymer performance and processing.

$\mathbf{MW_{ave}}$ and **MWD** significantly affect mechanical and physical properties of a plastic material, including tensile properties, creep and toughness. Polymer chains interact with one another both chemically and mechanically. The chains are entangled with one another, much like spaghetti strands in a bowl. As the strands get longer (a higher molecular weight) the amount of entanglement increases. Entanglement causes the chains to resist sliding over one another. The greater the entanglement, the more energy is required to slide the polymer chains across one another. In general we can say that higher molecular weights of a given polymer will have higher melting points and higher physical properties. As energy is applied to a polymer, as in adding heat, the added energy will increase vibrations at the molecular level, enabling the polymer chains to slide over one another causing the polymer to melt. Generally speaking, as molecular weights of a given polymer increase, physical properties will increase. Table 2.1 shows how changes in molecular weights generally will affect a number or properties.

The processing conditions required for a plastic material are greatly influenced by the **MWD**. Narrow **MWDs** will have a relatively sharp melting point, softening over a very narrow range of temperatures, while materials with a broad **MWD** will soften over a fairly broad range of temperatures and will not have a distinct melting point. This can be easily understood by visualizing what is happening at the molecular level. With a broad MWD plastic, as heat energy is applied molecular vibrations increase and the lower molecular weight (shorter) polymer chains are the first to start to slide over one another. The polymer will appear to soften, but the higher molecular weight chains will still be entangled and hold the polymer melt together. This characteristic is

**Table 2.1** General effect of MW on properties.

| As Molecular Weight Increases ↑ | Tensile Strength Increases ↑ |
|---|---|
| | Impact Strength Increases ↑ |
| | Solubility decreases ↓ |
| | Viscosity increases ↑ |
| | Melting Point Increases ↑ |

called melt strength. As heat continues to be applied, some of the longer chains will become unentangled and start to slide over one another. Eventually, all the chains will be able to move and we will have a complete melt. This point is not sharp or distinct. With a narrow MWD plastic, as heat is applied, all the chains will become free to move over one another over a very narrow range of temperatures. This narower melting point will allow the polymer melt to flow very easily, to be cooled and hardened very quickly, but will be difficult to use in processes requiring a high melt strength. We will discuss this more in Chapter 5 when we talk about individual processes.

## 2.2 Melt Flow Index

A relatively simple method to provide a general and relative characterization of how a given polymer will flow when melted is a test called the melt flow index (MFI), which is also sometimes called the melt flow rate (MFR).

In this test, a small sample of plastic is heated and forced through an orifice by a weighted piston. The results are recorded as the weight of material "extruded through the orifice" in a given amount of time (usually grams per minute at a given temperature). Typical values could be anywhere from 0.1 grams/minute for a very stiff, hard to flow material, up to 100 grams/minute for a very easy flowing material. A material with an MFI value of <1.0 is called a fractional melt material. The test methods and conditions are defined in ASTM D1238-Standard Method for Flow Rates of Thermoplastics by an Extrusion Plastometer. Because of its simplicity and repeatability this is a very common test used in industry, both for comparing one material to another and for quality control testing of a given material. Figure 2.4 shows a typical testing apparatus.

## 2.3 Molecular Structure of Polymers

How the polymer chains are packed together will also have a significant impact on the properties and processability of any given plastic. If the chains are randomly intertwined with one another the polymer is said to be amorphous. If the polymer molecule chains are packed together in regular repeating structural patterns the polymer is said to be crystalline. No polymer is completely crystalline, but is partially crystalline mixed in with some amorphous regions. The amount of the polymer

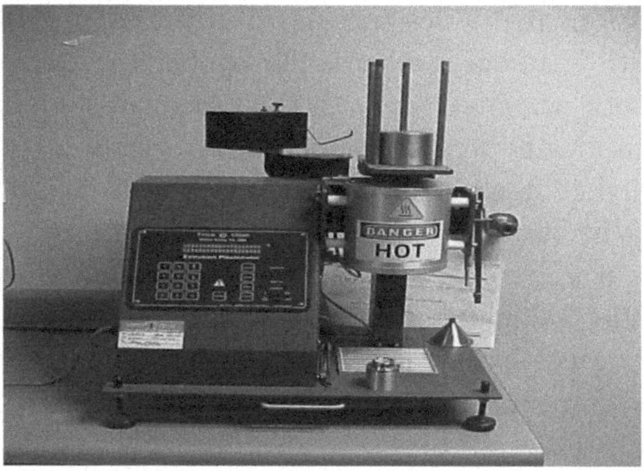


**Figure 2.4** Extrusion plastometer (melt index tester).

structure that is crystalline is referred to as the degree of crystallinity and is given as a %. The polymer with the highest % crystallinity is polytetrafluoroethylene (PTFE), which has a crystallinity of just over 70%. The degree of crystallinity is determined by the size and shape of the polymer chain, processing conditions and secondary bonding. If the polymer repeat unit is small and simple and the pendant groups are small, the polymer will be able to pack together and form crystalline regions. An example of this would be polyethylene, which consists of only carbon and hydrogen atoms. In the crystalline regions the polymer chains are folded over one another and packed tightly together. This close packing leads to additional secondary chemical bonding, which increases the strength of the polymer structure. As the molecules become longer and more complex and the pendant groups larger, this becomes less likely. An example of this would be polystyrene, which has a large benzene ring as a pendant group. The effects that a large atom or group of atoms have on polymer chains are called "steric effects."

The higher the density of a given polymer, the higher the crystallinity (the polymer is packed together more tightly). PTFE, the plastic material with the highest % crystallinity also has the highest density, 2.2 grams/cc. Ways to measure crystallinity include X-ray diffraction and infrared spectroscopy.

## 2.4 Thermal Properties of Plastics

Changes in temperature, sometimes surprisingly small, can have a significant effect on polymer properties and functional performance. One of the key decisions that a designer has to make when designing a product with a plastic material is what is the temperature range that the part **might** see. I once was involved with the design of a piping product for high temperature and pressure applications. We were very careful to consider the highest temperatures that the product might accidently be subjected to. What we overlooked was that if the product was shipped to Minnesota in February, it would also see extreme cold. The pipe became so brittle, it couldn't survive shipment.

The first concept that we will discuss is that of the melting point, $\mathbf{T_M}$. $\mathbf{T_M}$ is the temperature(s) at which a material changes from solid to liquid. It is the temperature where the molecular motion permits the polymer chains to slide over one another. At the $\mathbf{T_M}$ the polymer has a general liquid-like behavior.

Amorphous polymers soften and flow over a fairly broad temperature range. These materials do not have a distinct melting point and values for melt temperature are usually given as a range. As we discussed earlier, the broader the MWD for a given polymer, the larger will be the range of $\mathbf{T_M}$.

Crystalline polymers are made up of a mix of crystalline and amorphous regions. They will have a sharper melting point than an amorphous polymer. The higher the crystallinity, the narrower will be the range of melting temperatures. This sharp melting point is associated with the breaking of the secondary bonds mentioned previously. The amorphous regions will already be softened or melted, and when sufficient energy is applied to break the secondary bonds, the material will completely melt over a very small temperature range, sometimes as small as 1–2 °F.

At low temperatures polymers are solid and glasslike. This is said to be the glassy phase. Molecular motion in this phase is very limited. As we add heat to the polymer, molecular motion increases and the polymer chains start to move apart to accommodate this motion. This movement manifests itself as a change in dimension and is called thermal expansion. We measure this by the coefficient of linear thermal expansion (CLTE). It is the ratio of change in linear dimension per unit temperature change as compared to the original dimension. Dimensionally it is cm/cm per degree C (metric) or inches/inch per degree F (English). The standard test for this is ASTM D696. This is very important to the designer, especially with parts that must fit together like gears, because the part dimensions will change with temperature. The functional design fit must accommodate the full range of temperatures the parts will see. While both plastics and metals will change

**Table 2.2** Examples of CLTE for plastics and metals.

| Material | cm/cm/°C x $10^{-5}$ |
|---|---|
| **Steel** | **1.08** |
| **Copper** | **0.9** |
| **Aluminum** | **2.2** |
| **Polyimide** | **1.3** |
| **PVC (rigid)** | **2.7** |
| **Polycarbonate** | **3.6** |
| **Polypropylene** | **8.6** |
| **HD Polyethylene** | **11.0** |

dimensions with changes in temperature, plastics will generally change more significantly. Table 2.2 gives some comparative examples.

A way to reduce the CLTE of plastic materials is to add fillers and/or reinforcements which can greatly reduce the CLTE. For example, by adding glass reinforcement to polypropylene, we can improve the CLTE from $8.6 \times 10^{-5}$ cm/cm/°C to $3.2 \times 10^{-5}$ cm/cm/°C, a reduction of over $2^1/_2$ times. We will discuss fillers and reinforcements in more detail at the end of Chapter 3, Review of Plastics Materials.

Another very important thermal property of plastics is thermal conductivity. Thermal conductivity is the measure of the rate at which a material will conduct heat energy. Because plastics molecules move internally when heated, they tend to absorb heat energy and are generally poor conductors, but very good insulators. This is why plastics are widely used for thermal insulation in construction. Dimensionally, thermal conductivity is given in watts per meter kelvin (W/m K) (metric) or BTU/hr-ft F (English). The standard test for this is ASTM C177. Thermal conductivity is sometimes referred to as the K factor. Table 2.3 gives examples of thermal conductivity of plastics, metals and insulating materials.

While low thermal conductivity is great when insulation is a desired property, it can cause a challenge to the designer when the part must be able to remove heat. We will talk about this in detail in the chapter on enclosure design but, again, we can modify the thermal conductivity properties through the addition of various fillers.

**Table 2.3** Examples of thermal conductivity.

| Material | W/m K at 25 °C |
|---|---|
| Copper | 401 |
| Aluminum | 205 |
| Stainless Steel | 16 |
| HD Polyethylene | 0.42–0.51 |
| Polycarbonate | 0.19 |
| PVC (rigid) | 0.19 |
| Glass wool insulation | 0.19 |
| Polystyrene Foam (EPS) | 0.033 |
| Polyurethane Foam | 0.021 |

In addition to its importance in product design, the generally low thermal conductivity of plastics plays a significant factor in manufacturing plastic parts as they do not conduct applied heat well, causing difficulty in achieving a uniform melt during processing.

Perhaps the most important concept for the designer to grasp in understanding thermal effects on thermoplastic materials is that of glass transition temperature. As we discussed earlier, at low temperatures plastics are solid and glasslike. As temperature is increased, the amount of internal molecular motion increases. At some point this motion becomes significant and the molecules can move past one another. This can occur fairly suddenly, over a change of just a few degrees. Where this change occurs is called the glass transition temperature, $\mathbf{T_G}$. Below $\mathbf{T_G}$ a plastic is rigid and hard. Above the $\mathbf{T_G}$, a plastic is softer and more flexible. The plastic will have its maximum physical properties such as tensile and compressive strength below the $\mathbf{T_G}$, above it will be more flexible. The design implications are obvious. If you are designing a structural component that requires physical strength, you want the material to be below the $\mathbf{T_G}$ over the entire temperature range the product might see (think of a bridge support). Conversely, if you are designing a flexible shock absorbing washer, you want it to be above its $\mathbf{T_G}$ over the entire temperature range the product might see (think of a car tire).

Figure 2.5 shows how a plastic materials flexibility and softness increases as the temperature is increased for both an amorphous and a crystalline thermoplastic. In the amorphous plastic, the material stays

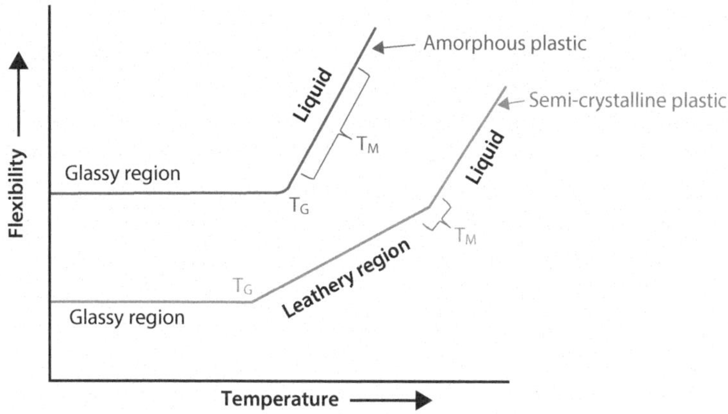


**Figure 2.5** Glass transition and melting point.

hard and glassy until enough heat energy is added to allow some of the molecules to start to move. Not all have enough energy yet to do so, so you get a softening of the material, but because some molecules are still relatively immobile, it is not yet a liquid state or a melt. As we increase temperatures, the material continues to soften and finally becomes liquid enough to melt and flow. The actual melting point is not sharp and well defined and is usually given as a range. From a design standpoint, the importance of the $\mathbf{T_G}$ is that at this point, the material will start to lose physical properties. From a processing standpoint, the range of $\mathbf{T_M}$ is the temperature you have to get the polymer to so it can flow and be processed.

Now let's take a look at the semicrystalline plastic (remember, no plastic is 100% crystalline, but a mix of crystalline and amorphous regions). Just as in the amorphous plastic, the material initially is hard and glassy. As we add heat energy the amorphous regions will start to soften. From the point the material passes through the glass transition and softens until the point that it melts is called the leathery region or rubber transition region. In this region the crystalline structure retains its properties until enough energy is added to break the secondary bonds holding the crystalline structure together. At this point, the molecules are free to move over one another and the material is both flexible and melted. This melting, $\mathbf{T_M}$, happens over a relatively narrow range of perhaps just a few degrees. An important difference between amorphous and semicrystalline plastics is that above the $\mathbf{T_G}$ semicrystalline plastics still have the ability to retain significant mechanical properties. This is because the crystalline melting point is higher than the glass transition point. Reduction of these properties, however, can still be substantial. The addition of a variety of reinforcements can substantially improve physical property retention at temperatures above $\mathbf{T_G}$.

To summarize, if you are designing a part for a structural application like a cell phone cover, you want strength and rigidity in the part. For this application, you would select a material that would be below its $\mathbf{T_G}$ for whatever use temperatures it might encounter (remember to consider shipping in hot sea containers!). Alternatively, if you are using a thermoplastic elastomer to design a valve seal, flexibility and resiliency is what is going to be important. You will want to select a material that will be above its $\mathbf{T_G}$ for whatever use temperatures it might encounter. Table 2.4 gives some examples of typical $\mathbf{T_M}$ and $\mathbf{T_G}$ for several materials. It is important to note that these values will vary greatly depending upon the specific grade of each polymer selected. It also has been found that $\mathbf{T_G}$ can vary depending on cooling rate [2].

With thermosetting plastics, the chains are locked together by crosslinked chemical bonds. Once crosslinked, these materials are hard at all temperatures. As the temperature is increased high enough, these bonds will start to break and the polymer will degrade and lose its properties. The melting point will be the temperature that will cause the polymer to flow, but as the chemical reaction goes forward and crosslinks are formed, the material will permanently harden.

**Table 2.4** Examples of $T_M$ and $T_G$ for selected plastics.

| **Material** | **TM** | | **TG** | |
|---|---|---|---|---|
| | °C | °F | °C | °F |
| Polypropylene (PP) | 163 | 334 | 5 | 41 |
| Polystyrene (PS) | 100 | 212 | 90 | 194 |
| Polyvinylidene dichloride (PVDC) | 172 | 277 | −15 | 5 |
| Acrylic (PMMA) | 160 | 320 | 100 | 212 |
| Polycarbonate (PC) | 266 | 510 | 150 | 300 |
| Nylon (PA) | 260 | 500 | 50 | 122 |
| ABS | 105 | 221 | 102 | 215 |
| Polyethylene Terephalate (PET) | 250 | 483 | 70 | 158 |
| Polyphenylene Sulfide (PPS) | 280 | 585 | 90 | 194 |

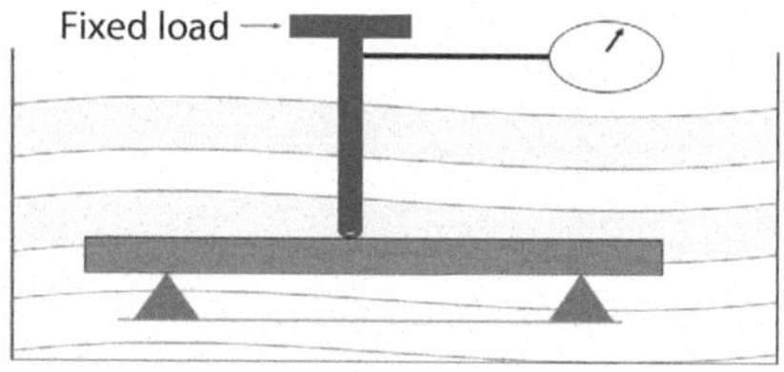


**Figure 2.6** Heat distortion temperature (HDT).

The degradation temperature ($\mathbf{T_D}$) is the temperature at which irreversible damage is done to the polymer. In thermosets, as mentioned above, it is the point at which the crosslinked bonds are broken. In thermoplastics, it is the point where sufficient energy has been applied to start to break the covalent bond in the polymer chains. In both cases, the polymer loses physical strength and cannot be repaired.

Heat distortion temperature (HDT) is a property that is commonly used to help select plastic materials to be used in a structural application at an elevated temperature. HDT or Deflection Temperature Under Load (ASTM D648) is a method of measuring the temperature required to produce a fixed deflection (0.010″) of a 5″ × 1/2″ × 1/4″ test bar. Figure 2.6 shows a schematic of the test apparatus used. Usually run at 264 psi or 66 psi, it is a good measure of what the temperature limitation will be for a plastic part under load. When comparing numbers for HDT from tables you need to be sure that the data you are comparing is based on the same load.

When a plastic is placed under load it will undergo a phenomenon called creep or cold flow. Creep is defined as increasing strain over time in the presence of constant stress [3]. The rate of creep for a plastic material is dependent on the applied stress, the temperature and the time that the stress is applied. Creep is the movement of the plastic under an applied load and its rate will increase with increasing temperatures. A plastic material subjected to a given load and temperature will deform over time in a very predictable way. Figure 2.7 shows a typical creep curve. As you can see, the part will initially deform at a higher rate but will continue to deform, although at a lower rate over time.

We will discuss the phenomenon of creep in more detail in Chapter 6, Structural Components.

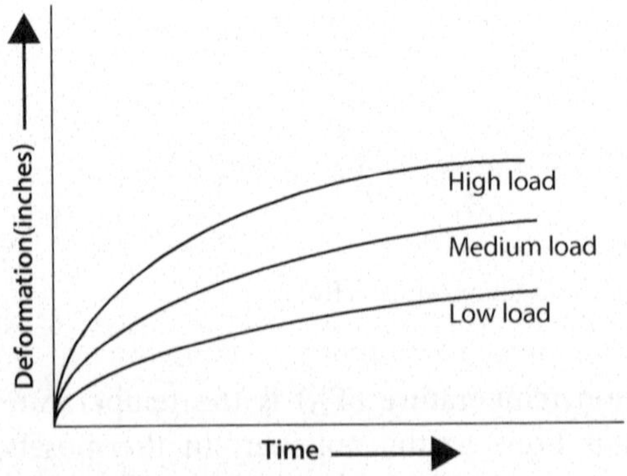


**Figure 2.7** Deformation over time or creep.

## 2.5 Physical Properties of Plastics

Next let's look at some physical properties of plastics. We first need to review a few basic definitions.

- Stress ($\sigma$) – Force divided by area (pounds per sq in or newtons per sq meter (pascal)
- Strain ($\varepsilon$) – The displacement of material under force (inches per inch). The change in length (L) divided by original length ($L_0$)
- Hooke's Law F/A = $\sigma$ = E $\varepsilon$, where E = Young's Modulus

A designer using plastic materials for the first time will find that, when viewed in the context of traditional materials, they behave in a unique and different way. Traditional materials are generally divided into two categories, elastic solids (such as iron, concrete and copper) and viscous fluids (such as oil, water and molten steel [4]. When a force is applied to an elastic solid, its shape will deform, but when the force is removed, the shape will recover. When a viscous fluid is exposed to a force it will move or flow. When the force is removed, the movement or flow will stop but it will not return to its original shape. Plastic materials behave in a manner that is a hybrid of an elastic solid and a viscous fluid. We call this viscoelastic or, sometimes, plastic behavior. Figure 2.8 illustrates the different ways that materials will respond to an applied force. In all three graphs in the illustration, we apply a constant force to the materials from time t = 0 to time t. At time t we remove the force.

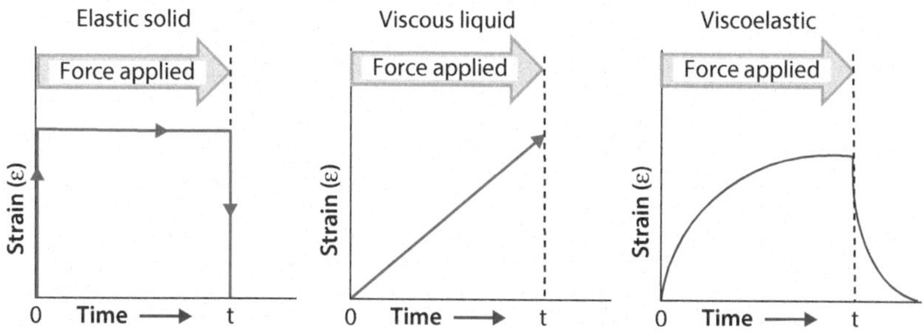


**Figure 2.8** Material response to an applied force.

The graph at the left shows what will happen with an elastic solid. When we apply the force, we will immediately deform (move) the material, shown by an immediate strain $\varepsilon$. The strain will remain constant as long as the load is applied. When we stop applying the force, the strain immediately returns to 0 and the deformation is recovered. Next, we look at what happens with a viscous liquid. When we apply the force, the material will start to deform, but unlike the elastic solid, the strain (and deformation) will continue to increase as long as the load is applied (think of oil being moved by a hydraulic pump). When the force is removed, the liquid will stay where it is and not recover its original shape. With a plastic or viscoelastic material, as we apply a force, the material will deform, and like a viscous liquid, the deformation and strain will continue to increase as long as the load is applied. When the load is removed, the plastic will recover its shape, but unlike an elastic solid, it will do so over a period of time. The time it takes to recover is called the relaxation time or the relaxation response.

Most elastic solids follow Hooke's law, as shown in Figure 2.9. Stress is proportional to strain in a linear manner within the usable strength range of the material. The slope of the linear area is the modulus of elasticity or Young's modulus (E).

Plastic materials will respond similarly over the linear range, but above it the viscoelastic nature causes them to behave a little differently. As you can see in Figure 2.10, as stress is applied to a plastic material, initially the strain will increase linearly. In this linear region we can calculate a modulus value. It should be pointed out that for most plastic materials this linear range is quite small. As the stress increases, a point is reached (the proportional limit) where the strain starts to increase at a greater than a proportional rate. This will continue until the elastic limit or yield point is reached. This is the point at which the

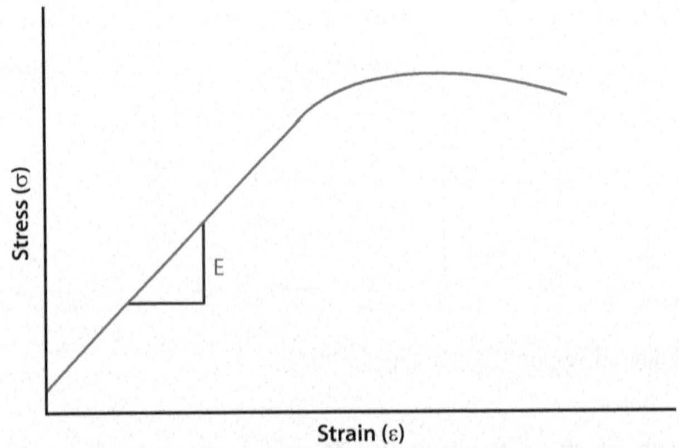


**Figure 2.9** Stress-strain relationships for an elastic solid.

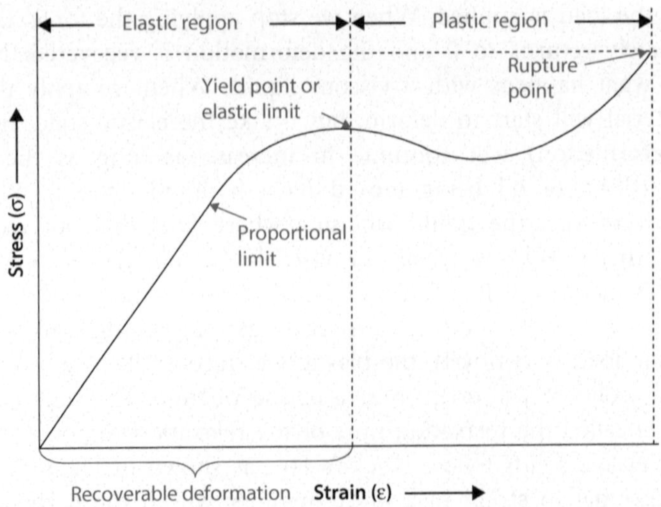


**Figure 2.10** Plastic stress-strain behavior.

deformation of the material is no longer recoverable. If the load is removed after this point, the material will not return to its original shape. At this point, at a molecular level, the polymer chains are starting to slide past one another. If more stress is applied, the part will rupture as the polymer chains are pulled apart and broken. This is the tensile strength of the material. As temperatures are increased, the rate of strain will increase, resulting in a lower tensile strength as temperatures increase. The HDT discussed earlier gives the designer a good indication of the sensitivity of strength to increasing temperatures. Tensile strengths for plastic materials are determined by ASTM standard D638. Flexural strength (strength in bending) is determined by ASTM D 790 and is given in psi or Mpa.

Hardness is the resistance of a material to surface indentation or deformation. Hardness in plastics materials can be measured in plastics similarly to how it is done in metals. Rockwell hardness (ASTM D785) is a common test used with harder materials. The "C" scale is typically used with plastics materials. The Shore Durometer test (ASTM D2240) is commonly used with softer materials. Both of these tests measure penetration of a rounded pin under load. The Barcol Hardness test (ASTM D1785) is used for very hard thermoset plastics and composite plastics. This test measures penetration of a sharp point into the test material. It is sometimes used to assess the degree of cure of a thermoset material.

Impact strength is a measure of a material's ability to absorb the energy of an impact event. The two most common tests used of interest are the Izod and Charpy impact test (ASTM D256). In the Izod test, a test bar is clamped at one end and held vertically. The bar is impacted by a weighted swing pendulum and the amount of energy required to break the sample is measured. The Charpy test is similar, except the bar is held horizontally and supported at both ends, and impacted in the center. Results are given in ft lbs per inch. Figure 2.11 shows a typical testing apparatus. In both tests there are test conditions for both a sample with a notch cut into it to provide a stress concentration and for an unnotched sample. The notched test enables achieving a failure in

**Figure 2.11** Impact strength tester.

ductile plastics that would not fail in an unnotched test. When comparing data it is important to be sure the test conditions are comparable.

Another impact test of importance is the Falling Dart impact test, also known as the Gardner impact test. It is a test which is commonly used to test the impact strength or toughness of plastic films and sheets. The test method, described by ASTM D1709 or ISO7765-1, involves clamping the plastic sample in a ring below a drop tower. A weighted dart is dropped from varying heights and the mean failure height, the point at which 50% of the samples tested will fail, is determined.

## 2.6 Electrical Properties

Plastics offer the designer a wide array of possible combinations of electrical properties and mechanical properties, from very rigid and hard to soft and flexible and from very insulative to conductive. While plastics are well known for their electrical insulative properties, they also can be used in electrostatic dissipative applications and conductive applications (see Figure 7.11 in Chapter 7 for more detail). There are a wide range of electrical properties that are used for a variety of reasons. Below we will discuss the most common.

The first property we will look at is resistivity. Resistivity is a measure of the resistance a material presents to the flow of an electrical charge. Surface resistivity is the measure of a material's ability to resist the flow of current along its surface when a voltage is applied to two surface-mounted electrodes [5]. ***Surface resistivity*** is measured in ohms. ***Volume resistivity*** is measured through the sample by electrodes on the top and bottom of the sample. Volume resistivity is measured in ohm-m (Ω/m) or ohm-cm, also called Ω/square. Both of these tests are described in ASTM D257.

***Dielectric strength*** is a measure of the breakdown voltage required to form a conductive path. The voltage/thickness equals the dielectric strength and is given in units of kV (kilovolts)/mm. This test is described in ASTM D149.

***Relative permittivity*** or Dielectric constant is a measure of how well a material will act as a dielectric. This test compares capacitance of the sample material to that of a vacuum (as a ratio). The higher the permittivity, the better the material will function as an insulator. This test is described in ASTM D150.

***Dielectric dissipation factor*** is the ratio of the electrical power dissipated in a material to the total power in the circuit [6].

Obviously, these properties would be of interest when designing parts that are in direct contact or in proximity to an electric current.

## 2.7 Flammability

In many design projects, the flammability of the materials being used needs to be considered very carefully. Some plastics burn readily, others are much more difficult to burn, and others will stop burning when the flame is removed (self-extinguishing). An equally important factor that must be considered is that of smoke generation, the amount of smoke that the material will produce when burned. Flammability is a very complex topic. Many industries have their own specific testing protocols that are used but there are several widely used methods that we will discuss here. The first is the UL 94 Flammability test. This test measures material burning characteristics. A test bar is clamped either horizontally (horizontal burn) or vertically (vertical burn) and a flame is brought into contact with the sample. The flame is removed and the results on the sample bar are characterized. This test is used in appliances and business machines.

Another widely used test is the ASTM E-84 (UL723) test (also called the Steiner tunnel test). This test measures flame spread and smoke generation in a controlled environment and is good for comparing relative material performance. This test appears in a number of building codes. Figure 2.12 illustrates how the flame spread is calculated in this test.

Figure 2.13 illustrates how the test measures smoke generation.

Another widely used test is ASTM D2863 – Limiting Oxygen Index. This test measures the minimum oxygen concentration required for

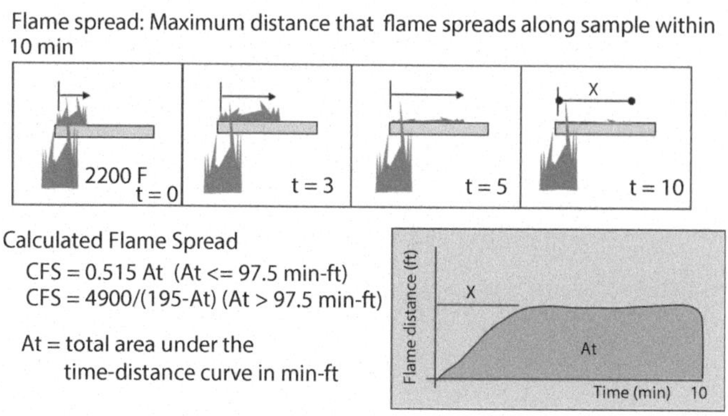


**Figure 2.12** Flame Spread – ASTM E-84 (UL-723).

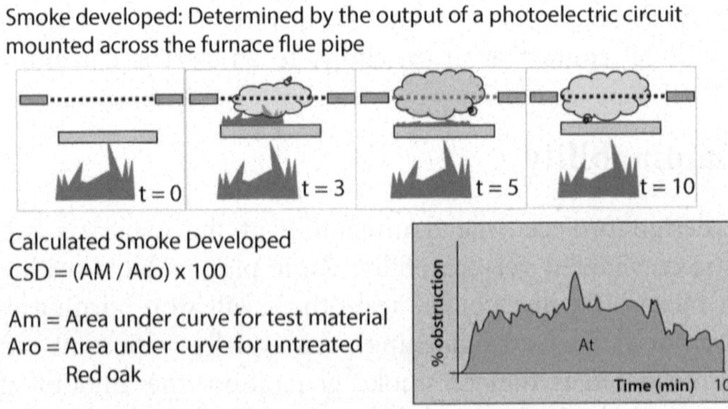


**Figure 2.13** Smoke Generation – ASTM E-84 (UL-723).

combustion. The specimen is positioned vertically in a transparent test column and a mixture of oxygen and nitrogen is forced upward through the column. The specimen is ignited at the top. The oxygen concentration is adjusted until the specimen just supports combustion. The concentration reported is the volume percent.

The National Bureau of Standards smoke generation test as described in ASTM E662 is another test used to characterize smoke generation.

# 3

# Overview of Plastics Materials

There are many different plastics materials available to the designer, all similar in some ways, but each offering some unique properties that make them the preferred material of choice for specific applications. These materials are being continually improved by the resin manufacturers to enhance their properties and expand their range of applications and new materials are being continually developed and introduced that allow plastic materials to be used in new and innovative applications. The discussion that follows will attempt to give a current overview of the major materials that are now in use.

Plastics are generally classified into two basic families, thermoplastics and thermosets. Within each of these families we further classify materials as commodity plastics, engineering plastics and high performance polymers. Commodity plastics are common, relatively low cost polymers that are essentially interchangeable in use from manufacturer to manufacturer. They also are relatively on the low to mid end of both thermal and physical properties. Engineering plastics are those materials that offer superior performance in one or more properties. The high performance polymers, as the name implies, are the latest very high end

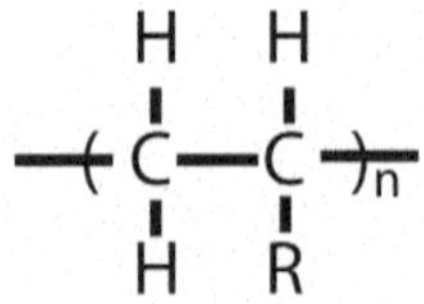

**Figure 3.1** Commodity plastics chemical formula.

performance polymers. This category has steadily changed over the years and materials such as polycarbonate, which was once hailed as a high performance polymer with exceptional properties, now falls in the midrange of engineering thermoplastics.

First, let's take a look at commodity thermoplastics. There are four basic commodity plastics, polyethylene, polypropylene, polystyrene and polyvinyl chloride. They are all made by addition polymerization, which is a very cost-effective method of polymerization that readily lends itself to very large-scale production. These materials can all be represented by the following chemical formula in Figure 3.1, where R represents various functional groups:

If the R = H (hydrogen) we have the polymer polyethylene, if the R = $CH_3$ (a methyl group) we have polypropylene, if R = $C_6H_6$ (a benzene ring with alternating single and double bonds commonly shown by the 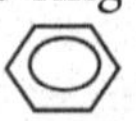 shorthand) we have polystyrene, and if R = Cl (chlorine) we have polyvinyl chloride.

These basic commodity plastics each have a family of copolymers and blends, some of which have some exceptional properties that are not commodity plastics per se but will be discussed with each of these commodity plastics.

## 3.1 Polyethylene

Polyethylene (PE) is the simplest of all polymers consisting only of hydrogen and carbon atoms. It is manufactured by the addition polymerization of ethylene ($H_2C = CH_2$) into the polymer (see Figure 3.2).

Polyethylenes have a waxy feel similar to that of paraffin wax, their shorter chain relatives. The basic polymer structure of PE is what determines its properties. The polymer is compact and tightly packed together, making it insensitive to solvents, and the similar

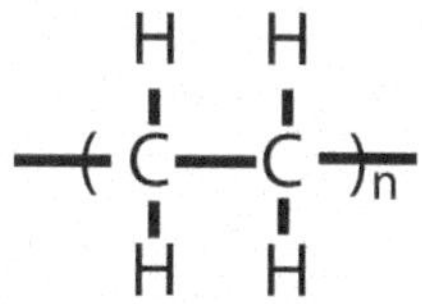

**Figure 3.2** Polyethylene polymer.

electronegativity of the carbon and hydrogen results in low polarity, making it a very good electrical insulator.

There are three basic types of polyethylene. The first is called low-density polyethylene (LDPE). It is produced by a high-temperature, high-pressure polymerization process (50,000 psi, 300 °C) which yields a product with a specific gravity (density) 0.910–0.925 g/cm$^3$, has good flexibility and toughness, and is easy to process. This was the original process for polyethylene and is now found only in older plants. It is used in many diverse applications, including packaging film and bags, toys and squeeze bottles.

The second major type of polyethylene is called linear low-density polyethylene (LLDPE). Technical advances in catalysts and reactor chemistry allowed producing a branched polyethylene product similar to low-density polyethylene, but at much lower temperatures and pressures. For safety and cost reasons this is the process of choice and all new plants are built to this technology. The process yields a specific gravity (density) of 0.918–0.940 g/cm$^3$ and properties similar to LDPE, but a little higher stiffness is possible. The applications are similar to LDPE, but the slightly better strength has made this the material of choice for trash bags. The majority of LLDPE resins are manufactured with Zeigler-Natta catalyst systems but about 25% are produced using metallocene catalyst technology. Metallocenes produce resins with greater clarity and dart impact strength, which is important for plastic films, but are a little more difficult to process. Also, plastic resins are more expensive to produce using metallocenes [7].

The third major type is high-density polyethylene (HDPE). Again, using ethylene as a feedstock, reacting with a Zeigler-Natta catalyst produces long un-branched chains at low temperatures and pressures, yielding a denser product (specific gravity [density] 0.935–0.960 g/cm$^3$) with improved stiffness and mechanical properties and excellent chemical and electrical resistance. It is widely used for water and gas pipe, storage containers, pallets and trash bins.

There are also a few non-commodity variants of PE that should be mentioned. The first is a crosslinked polyethylene or PEX. It is produced by inducing crosslinks between the PE chains. Currently there are three processes in commercial use. PEXa is a high pressure process (Engel Process) where peroxide is mixed with the PE prior to extrusion and the crosslinks are formed as the material is extruded. PEXb is the silane or moisture cure process. In this process, the crosslinking takes place after the material is extruded. The reaction is driven by moisture and temperature due to exposure to hot water or steam. Lastly, in the PEXc process, normal HDPE is extruded and crosslinks are induced by exposure to a high energy electron beam. This material has found widespread use for water distribution in residential housing. This product is made by crosslinking regular polyethylene with an additional process.

Ultrahigh molecular weight polyethylene (UHMWPE) is an extremely high molecular weight PE with MWs over 10 times greater than normal PE. It has excellent low temperature toughness and abrasion resistance properties. It is extremely tough to process making it usable only where its superior properties are absolutely required. Applications include liners for coal cars, slurry conveying, ice rink liners, gears and applications that need high resistance to wear.

There are also some important copolymers of polyethylene. A copolymer is a polymer that is made by polymerizing two or more different monomers together. The monomers can be joined together in several different ways. Figure 3.3 shows some of the possibilities.

Ethylene vinyl acetate copolymer is a copolymer of ethylene and vinyl acetate.

The vinyl acetate group provides a lot of steric interactions and the polymer is highly polar, making it able to bond well with other materials. It is used in hot melt adhesives, plastic cling wraps and in flexible tubing.

```
( X–Y–Y–X–X–X–Y–X–Y )n     Random copolymer
( X–Y–X–Y–X–Y–X–Y )n       Alternating copolymer
( X–X–X–Y–Y–Y–X–X–X )n     Block copolymer
( X–X–X–X–X–X–X–X–X )n     Graft copolymer
            |
            Y
            |
            Y
```

**Figure 3.3** Examples of copolymers made of two monomers, monomer X and monomer Y.

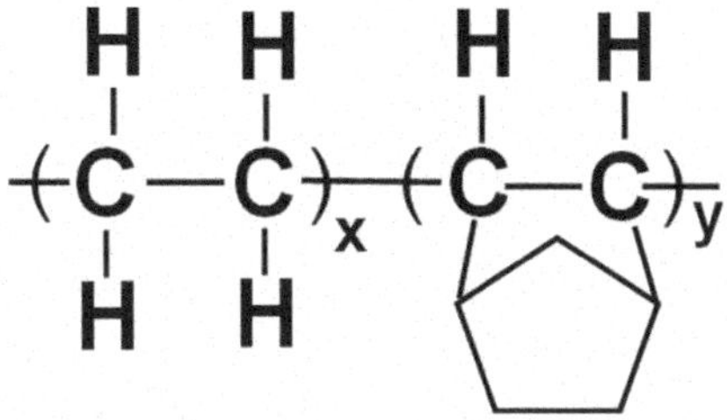

**Figure 3.4** Cyclic olefin copolymer.

Ethylene acrylic acid copolymer is a very interesting copolymer of ethylene and acrylic acid. The acrylic acid can give up a hydrogen ion which in the presence of a base can form an ionic bond. This, along with the polar pendant acrylic acid group creates a material that is very tough with good water resistance. The material is widely used for golf ball covers and adhesives and may be familiar to the reader by the trade name DuPont Surlyn ™ Ionomers.

The last PE copolymer that will be discussed is a cyclic olefin copolymer (COC) (Figure 3.4). This is a relatively new copolymer invented in the late 1990s. If we were not including this material as part of the PE family, its enhanced properties would qualify it as an engineering resin.

The unique structure including the large pendant pentene ring provides some unique properties. This is an amorphous transparent copolymer with extremely high transparency (91% light transmission), heat resistance and a very low density (1.02). It also has excellent chemical resistance and is sterilizable. Its uses include lenses, optical storage, and medical devices.

Polyethylene is the highest volume plastic used both in North America and worldwide. US demand in 2013 was over 38 billion pounds and total demand was over 179 billion pounds worldwide. The chart in Figure 3.5 shows how that is divided among the three major types of polyethylene.

## 3.2 Polypropylene

Polypropylene is manufactured by the addition polymerization of propylene ($CH_2CHCH_3$).

Polypropylene (PP) is structurally similar to PE with the exception of a methyl ($CH_3$) group being substituted for one of the hydrogens (see Figure 3.6). This larger side group provides some steric hindrance which

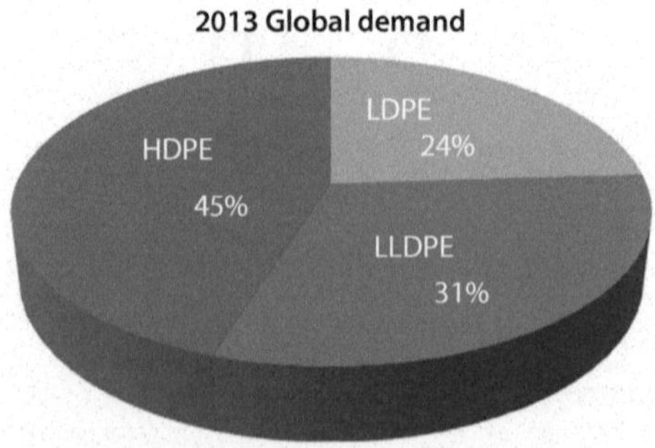


**Figure 3.5** Global demand for polyethylene in 2013.

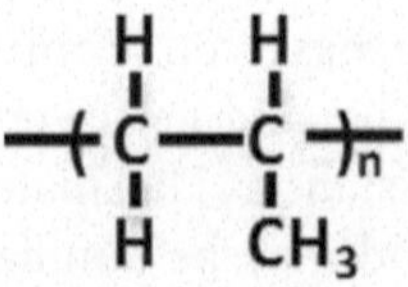

**Figure 3.6** Polypropylene.

gives the polymer higher heat resistance, toughness and strength compared to PE. It is also stiffer and more abrasion resistant and slightly more expensive. It is used for battery cases, low cost gears and sliding components, dishwasher safe food containers, toys and low cost furniture. Its superior toughness and long, entangled chains make this an excellent choice when producing parts with an integral (sometimes called "living") hinge. The use of Ziegler-Natta or metallocene catalysts allows producing a polymer with a regular repeating spatial structure. This produces a highly crystalline structure, which is the most common commercial configuration. The other is a terpolymer.

There are a couple of PP copolymers that are of commercial importance. One is an ethylene propylene copolymer (EPM) and the other (made with three monomers) is called EPDM. EPDM is made of ethylene, propylene and a material called a diene (a material with a carbon-carbon double bond). Both of these materials are above their $T_G$ well below room temperature and exhibit very high elongation. They are in a class of materials called thermoplastic elastomers or thermoplastic rubbers. They are widely used in "o" rings, gaskets, seals and flexible coatings.

Polypropylene volumes are over 16.5 billion lbs in North America and 90 billion lbs worldwide.

## 3.3 Polystyrene

Polystyrene is another addition polymerization product which is made from the monomer vinyl benzene. The polymer structure is shown in Figure 3.7.

The very large aromatic ring pendant (the drawing is not to scale) prevents crystallization in the polymer and increases the tensile strength. This large pendent group prevents the chains from packing together and results in a polymer that is almost 100% amorphous and therefore quite transparent. In addition to the excellent clarity, polystyrene is known for its tensile strength and resistance to water. It does have poor resistance to most solvents and although strong can be quite brittle. The material can be compounded with a variety of additives to improve its impact strength as well as weatherability and flammability. Polystyrene is used in a wide range of applications such as packaging, cups and appliances. We can make PS into a foam by using an inert gas. This material has a low thermal conductivity, high strength-to-weight ratio and will float, which lends itself to use as insulation, coffee cups, ice chests and pool toys.

A widely used commercial alloy of PS is made by dissolving butadiene rubber in styrene monomer and then polymerizing the styrene. This yields a butadiene rubber with PS side chains. This copolymer is then mixed with more PS, creating an alloy with up to 30% rubber content called high impact polystyrene or HIPS. (An alloy is the combination of two polymer melts into a single homogeneous phase. The melts must be miscible. This would be analogous to a metal alloy.) The alloying process improves the impact strength of HIPS to over seven times that of PS, but the rubber content results in a softer material that is no

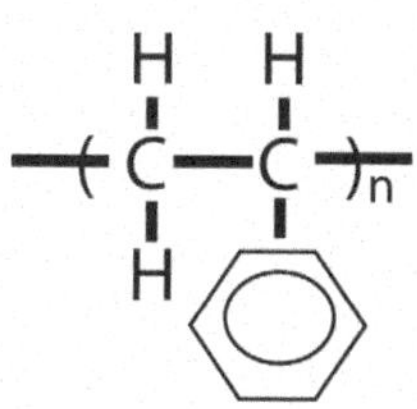

**Figure 3.7** Polystyrene.

longer transparent. HIPS is used in toys, appliances and packaging applications.

By copolymerizing styrene with acrylonitrile, we can produce a copolymer called styrene acrylonitrile or SAN. This material has improved chemical resistance, HDT, and tensile strength when compared to PS. It is still transparent but has a slight yellowish cast. Sometimes a colorant is added to give a more aesthetically appealing blue cast to the material. SAN is used for packaging where the improved chemical resistance is needed, in dishwasher safe containers and bottles.

Another widely used copolymer is Acrylonitrile Butadiene Styrene or ABS. ABS is a terpolymer (three monomers) whose properties can be adjusted by varying the proportions of acrylonitrile, butadiene and styrene. This gives a material that has excellent strength and toughness, good colorability and is very easy to process. Although generally opaque, there are transparent grades available. It is used in a wide variety of applications, including appliance cases and covers, tool housings, telephones, helmets and pipe.

Acrylic styrene acrylonitrile (ASA) is a terpolymer similar to ABS but is characterized by outstanding outdoor weatherability, UV stability, colorability and toughness. The improved weatherability is due to the acrylic elastomer. Applications include building and construction, recreation and automotive markets.

Polystyrene volumes are over 5.5 billion lbs in North America and 23 billion lbs worldwide. The chart in Figure 3.8 shows the major markets that polystyrene and polystyrene copolymers are used in.

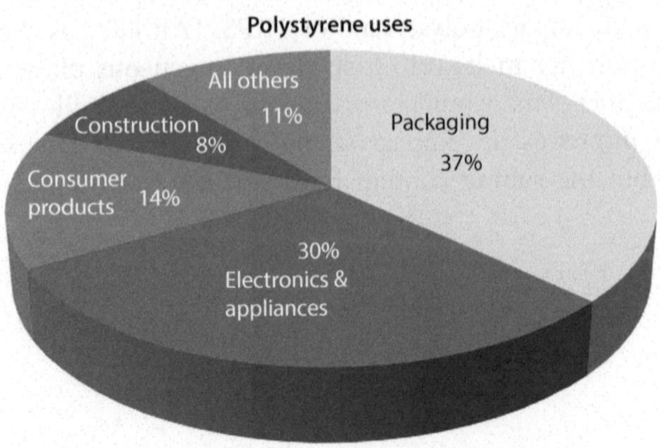


**Figure 3.8** Major markets for polystyrene and polystyrene copolymers.

## 3.4 Polyvinyl Chloride

Polyvinyl Chloride is made by the addition polymerization of vinyl chloride ($CH_2CHCl$) to produce an amorphous polymer as shown in Figure 3.9.

This is a relatively simple structure, very similar to polyethylene, with the substitution of a chlorine atom for one of the hydrogen atoms. The presence of this chlorine atom is what gives PVC some of its unique properties, such as excellent flammability resistance, but also is a primary influence on the difficulty of processing this polymer. The chlorine atom is about 2–1/2 times bigger than the hydrogen atoms. This creates a large steric effect that influences how the polymer melts. A large amount of energy has to be applied to allow the large chlorine side groups to slide over one another as the polymer melts. This energy is close to the amount of the bond energy holding the molecule together and some of these bonds can break, which results in hydrogen chloride (HCL) being formed, which further stimulates decomposition. To enable the processing of PVC, heat stabilizers which capture the HCL and processing aids which help the polymer chains slide over one another must be added.

Unmodified PVC (except for the additives to allow processing) is called rigid PVC or sometimes UPVC (unplasticized PVC). PVC can also be modified with a plasticizer to produce a flexible product called plasticized PVC or sometimes just vinyl. A plasticizer will enter the polymer structure and cause some swelling and disentanglement of the molecules [8].

Rigid PVC is a very stiff and tough material that has excellent flame resistance (the material is classified as self-extinguishing), excellent weatherability, and good chemical resistance to acids and bases. It is susceptible to solvents, which does allow it to be easily solvent welded together. This is a real advantage in the joining together of pipe and

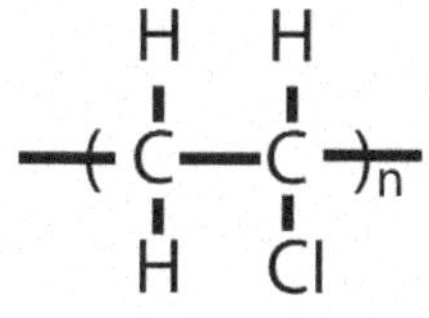


**Figure 3.9** Polyvinyl Chloride.

fittings. Applications for rigid PVC include pipe, pipe fittings, valves, siding, window profiles, decking, bottles, gutters, fencing and packaging.

Plasticized PVC is very flexible and unlike rigid PVC is above its $T_G$ at room temperature. It has a higher impact strength but lower tensile strength and HDT than rigid PVC. Flexible vinyls are processed similarly to rigid vinyls, but at lower temperatures. Flexible vinyls are used for bottles, floor mats, shrink wrap, medical and food storage bags and flexible tubing (*Tygon* ™).

Another interesting category of vinyl materials are plastisols. Plastisols are a vinyl that has been put into solution by immersing in a solvent. In this liquid form the material can be used in spray or dip coating applications used to make products such as vinyl gloves and tool handles.

There are a two interesting copolymers of PVC that need to be mentioned. The first is a PVC-polyvinyl acetate copolymer. This is the material used in vinyl flooring. The second is polyvinyl butyral. This is made by polymerizing polyvinyl alcohol (PVA) with butyraldehyde, yielding a tough, flexible material used as the inner layer of safety glass and as the back glass in thin-film solar modules.

There are also two other materials of commercial importance that are in the PVC family that will be discussed here. The first is chlorinated polyvinyl chloride (CPVC), which has two chlorine atoms, each attached to a different carbon atom. It is usually made by the post chlorination of PVC. This yields a 30% improvement in use temperature over rigid PVC as well as improved chemical resistance and flammability resistance. The downside of the additional chlorine is that the material is very difficult to process. CPVC is widely used as residential hot water piping, chemical processing pipe and in transportation applications.

The second is polyvinylidene chloride (PVDC). This material also has two chlorine atoms, but they both or on the same carbon atom, which yields very different properties from CPVC. PVDC is very tough and has excellent barrier properties. It is widely used as a shrink film and food wrap (Dow Saran®).

PVC volumes are over 15 billion lbs in North America and 83 billion lbs worldwide, making it the second largest volume plastic used today. The chart in Figure 3.10 shows the major markets that polyvinyl chloride is used in.

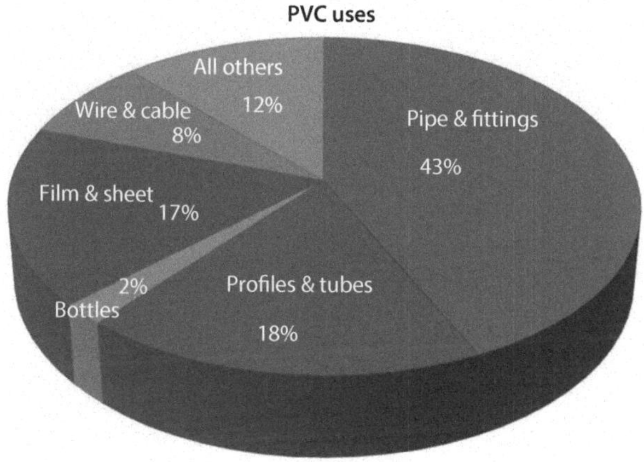


**Figure 3.10** Major markets for polyvinyl chloride.

## 3.5 Engineering Plastics

Engineering plastics offer the designer a range of materials that offer one or more properties with superior performance. These properties might include dimensional stability over a broad temperature range, high strength, improved stiffness and impact resistance. This combination of properties that each of these materials possess deliver a cost/performance advantage that make them the material of choice for many different applications. Because of the usually more complex process to produce these materials, more expensive feedstocks, and generally lower volumes, engineering plastics are a bit more pricey than the commodity plastics, and in some cases significantly more expensive.

### 3.5.1 Cellulosics

When John Wesley Hyatt developed celluloid back in 1868 he probably didn't realize that he just developed the first family of engineering thermoplastics. They qualify as an engineering plastic because of their excellent impact strength, environmental stress crack resistance, hardness, gloss, and colorability, as well as their unusual "warm" feel which lends itself to use in skin contact applications like glasses, hearing aids, and cosmetic cases, among others (see Figure 3.11). Cellulosics, when modified with additives to improve processing, are easily formed with standard manufacturing techniques such as molding and extrusion.

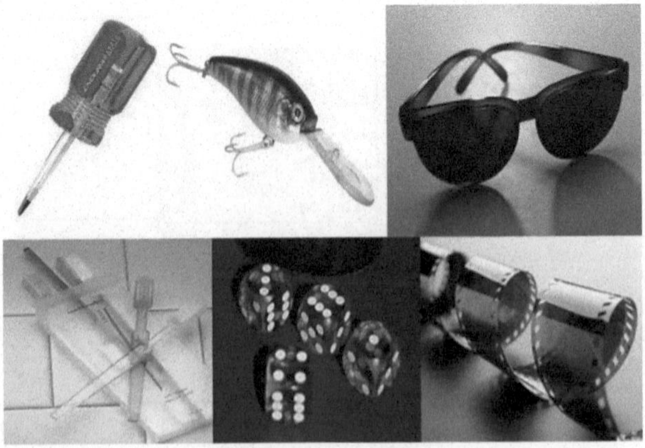

**Figure 3.11** Cellulose applications.

Cellulosics were the first plastics to be made from a renewable resource, wood. Wood pulp is treated chemically to make alpha cellulose fiber. The fiber is then treated with acids and anhydrides to make cellulose esters. The finished esters are a fine powder that is compounded with plasticizers and other additives to make the cellulosic plastic compound [9].

Commercial cellulosics are now made up of three primary materials. Cellulose Acetate (CA) possesses good gloss characteristics and clarity as well as excellent impact strength. It is used in applications such as tool handles, pen bodies, packaging (cellophane), filter mediums, film, and face shields.

Cellulose Acetate Butyrate (CAB) exhibits excellent impact strength. It is used for blister packaging, pen barrels, tool handles, and covers. Because of the butyric acid in it, the material can have a slightly unpleasant odor.

Cellulose Propionate (CP) has good impact strength and gloss and also has grades which carry FDA approval. It is used for toothbrush and hairbrush handles, ophthalmic frames, sunglasses, and packaging.

### 3.5.2 Polymethyl Methacrylate (Acrylic)

Polymethyl methacrylate (PMMA) (Figure 3.12), commonly called acrylic, is manufactured by the addition polymerization of methyl methacrylate.

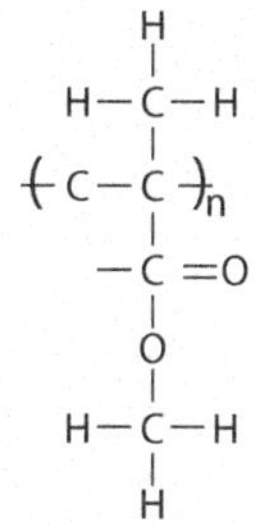

**Figure 3.12** Polymethyl methacrylate.

Acrylics have the best light transmission of any thermoplastic (92%) which lends it to wide use in a number of optical applications. It also possesses low sensitivity to oxidation and UV light, which gives it outstanding weatherability properties. The material can be readily colored and is available in a wide range of transparent, translucent and opaque colors. Applications include high quality lenses and optics, outdoor signs and skylights. The material is easily machineable and formable and finds wide use with hobbyists. In this market, it is commonly known by a variety of trade names, including Plexiglas® (trademark of Arkema) and Lucite® (trademark of Lucite International).

Acrylic can also be cast out of a solution in its own monomer (called acrylic syrup) and hardened by a curing agent and/or UV light. This material is used to cast clear sheets and rods, to make embedments, and with fillers added used to cast sinks, bathtubs and countertops. Acrylics are also used as blending and alloying resins for styrene to produce materials like ABS and SAN.

Figure 3.13 shows a landmark sign in Boston, MA, made of acrylic.

### 3.5.3 Polycarbonates

When polycarbonates were introduced in the late 1950s they were hailed as a breakthrough high performance polymer. The material had outstanding impact strength, was very dimensionally stable and had a continuous use temperature of 275 °F. Thermal properties like this had previously only been available in the more difficult to process thermosets.

The polar carbonate groups (C = O) allow hydrogen bonds to form between pendant groups (see Figure 3.14), which yields the material's excellent impact toughness. This polar nature makes it sensitive to solvents and makes the material hydroscopic. The polymer is amorphous

**Figure 3.13** Acrylic landmark sign.

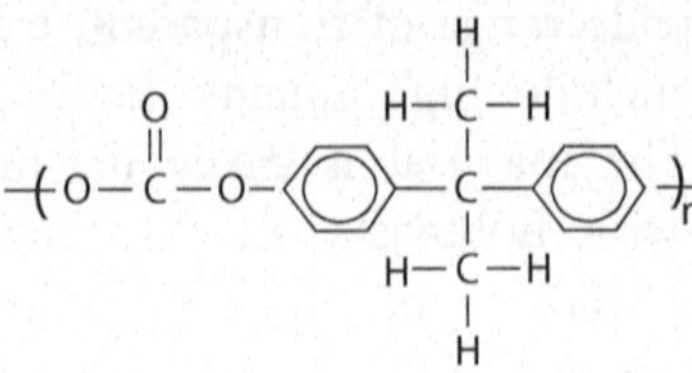

**Figure 3.14** Polycarbonate.

and displays excellent transparency. PC is sensitive to UV light and needs protection for outdoor applications. Concerns about the toxicity of bisphenol A, which is used in the manufacture of the resin, have limited its application in food service applications.

A recent introduction to the polycarbonate family is a copolymer of PC and polysiloxane. This copolymer has outstanding low temperature impact and improved overall impact strength.

Polycarbonate's outstanding impact performance makes it the material of choice for safety helmets and safety glasses. It is also used in lenses, tool housings, hair dryers, camera housings, cell phone cases, business machine and computer housings.

A widely used alloy of polycarbonate is polycarbonate/ABS (PC-ABS). This material molds much more easily than PC and exhibits good low temperature toughness and dimensional stability. Its strength, stiffness and heat resistance is between that of PC and ABS. It has a

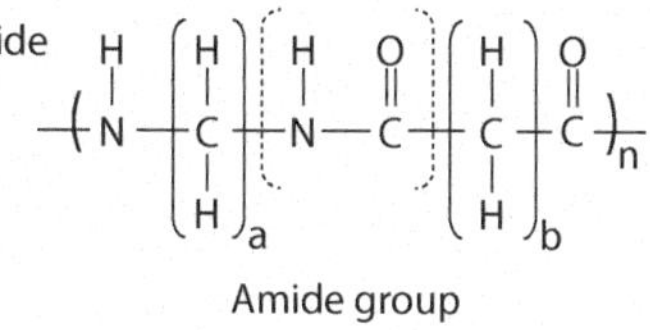


**Figure 3.15** General formula for nylons.

**Table 3.1** Naming of nylons.

| Monomer A | No. C's | Monomer B | No C's | Nylon Type |
|---|---|---|---|---|
| Hexamethylene diamine | 6 | adibic acid | 9 | Nylon 6/6 |
| Hexamethylene diamine | 6 | sebacic acid | 10 | Nylon 6/10 |
| Hexamethylene diamine | 6 | undecanoic acid | 11 | Nylon 6/11 |

lower cost than PC and is used in appliances, business and electronic equipment, automobile applications and for 3D printing.

### 3.5.4 Polyamides (Nylon)

When Dr. Carothers developed nylon through the condensation polymerization of an organic di-amine and an organic di-acid, it was a breakthrough in polymer chemistry technology in that he engineered a polymer with specific properties in mind. His discovery has led to a family of nylon materials with a broad range of properties. The general formula for nylons is shown in Figure 3.15.

By varying the number of carbon atoms in the organic di-amine (monomer "a" above) and in the organic di-acids (monomer "b" above) nylons with different properties can be produced. The nylon is named by the number of carbon atoms on each of the monomers. Table 3.1 shows how this naming convention works. For example, if a 6 carbon di-amine (hexamethylene di-amine) is used and a 10 carbon organic acid (sebacic acid) is reacted with it, the nylon produced will be called Nylon 6/10, noting the carbon atoms in the di-amine and the di-acid respectively.

An exception to this convention is when a single monomer ring compound e-caprolactam is used. When the ring is opened it has both an acid and an amine group which polymerizes to what is called nylon 6.

The nylon polymer is quite polar. This polarity induces secondary bonds which lead to tight packing of the polymer chains and high crystallinity.

Nylons in general are flexible, tough, strong plastics with a low coefficient of friction. They are hydroscopic and the absorption of water can affect properties and dimensions negatively. They accept fillers and reinforcements well, which are frequently added to further enhance properties. As mentioned previously, nylon is a large family of materials. Below we will discuss some of the more important ones.

Nylon 6/6 was the original nylon and is still the workhorse of this family of materials. It is flexible, tough and is widely used in industrial and carpet fibers, coatings, brush bristles and general moldings.

The special case nylon 6 generally has slightly lower performance than 6/6 but its lower price makes it attractive for a number of applications.

Nylon 6/10 is more flexible and tough with lower water absorption and better dimensional stability than nylon 6/6. It is used to produce gears, cams, zippers and insert fittings.

Nylon 6/11 is a "green" polymer. The 11 carbon acid used to produce the polymer comes from castor beans, a renewable feedstock. It has a higher tensile strength and lower water absorption than nylon 6/10 and is a little stiffer. It is used in gas pipe and fittings, impact resistant coatings and brake tubing. It is more expensive than nylon 6/6 or 6/10.

Nylon 6/12 is similar in properties and applications to nylon 6/11 and can be a little easier to process.

Nylon 4/6 has good thermal properties, offering a higher heat deflection temperature than other nylons. It is commonly used with glass and carbon fiber fillers which further enhance its ability to handle high temperatures. It is used in automotive under the hood applications, electrical components and appliances.

### 3.5.5 Polyoxymethylene (Acetal)

Polyoxymethylene (POM), commonly called acetal, has a high crystallinity which yields excellent strength, stiffness, barrier properties and solvent resistance. The diagram in Figure 3.16 shows the relatively simple structure which helps the polymer chains to pack tightly together.

It has better resistance to water absorption than nylon but is notch sensitive, which means that it is susceptible to cracking at sharp corners, which must be avoided in any design. Acetals come in two forms, a homopolymer (such as DuPont trade name Delrin®) which exhibits

POM or Acetal

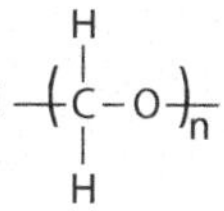

**Figure 3.16** Polyoxymethylene (Acetal).

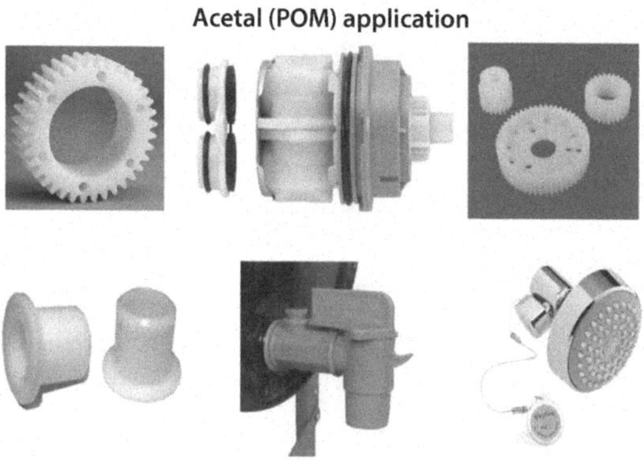


**Figure 3.17** Acetal (POM) applications.

higher tensile strength, modulus, hardness, and impact strength and a copolymer (such as Ticona trade name Celcon®) which has improved thermal stability, more flexibility and elongation. Ease of degradation makes processing this polymer critical and difficult. The homopolymer is slightly harder to process than copolymer. It is very important to process this polymer properly to insure proper crystallization to achieve optimum part properties.

Because of its excellent resistance to water, low coefficient of friction and toughness, acetal is widely used in residential faucet valve bodies. It is also used in shower heads, toilet tanks, gears, zippers and chain links (Figure 3.17).

### 3.5.6 Thermoplastic Polyesters

Thermoplastic polyesters are comprised of two basic materials, polyethylene terephthalate (PET) and polybutylene terephthalate (PBT) (see Figure 3.18).

Thermoplastic polyesters have good barrier properties, good water resistance, tensile strength and impact strength. PET is used in packaging and magnetic tape and is also used extensively as a stain resistant,

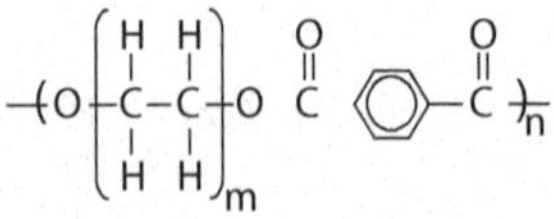


**Figure 3.18** Basic materials comprising thermoplastic polyesters.

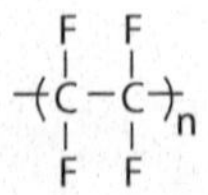

**Figure 3.19** Polytetrafluoroethylene.

wrinkle resistant fiber in clothing and as a hollow fiber for thermal insulation. Amorphous PET (APET) resins are used in soft drink containers (low permeability to carbon dioxide).

PBT processes a little more easily than PET, although it generally has slightly lower physical properties. It is used in applications such as gears, pulleys, cams, bearings and electrical components.

PETG is an interesting copolymer made from a modified glycol amorphous PET. The copolymer inhibits the material from crystallizing, giving it excellent clarity. It has excellent impact strength, but is susceptible to UV attack. It is used widely for soap dispensing bottles and packaging.

### 3.5.7 Fluoropolymers

Fluoropolymers are a family of plastics that were initially discovered by accident. In 1938, DuPont chemist, Dr. Roy J. Plunkett, was working with refrigerants. Upon checking a gas cylinder of compressed tetrafluoroethylene, he discovered that it had polymerized into a white, waxy solid to form polytetrafluoroethylene (PTFE). If you look at the polymer structure (Figure 3.19), you will see that it has a relatively simple structure, much like polyethylene, with fluorine atoms substituted for all the hydrogen atoms.

The presence of fluorine in the polymer gives these materials some very unique physical and chemical properties. The highly electronegative fluorine holds the electrons tightly and forms very stable bonds, which make it very chemically inert. The tendency not to bond with anything else gives fluoropolymers their trademark nonstick properties. You will, of course, recognize PTFE's trade name Teflon® (trademark of E. I.

DuPont De Nemours and Co.). The simple structure allows the long and straight molecules to tightly pack together, yielding the highest densities of any plastic (>2.2 g/cc). Also, PTFE is >70% crystalline, the highest crystallinity of any plastic. Fluoropolymers have excellent solvent resistance, excellent electrical resistivity and a very low coefficient of friction. PTFE is very difficult to process. Its melting point is close to its decomposition point. Conventional manufacturing techniques do not work. The material can be processed by sintering like a ceramic or by ram extrusion. Most parts are manufactured by machining from bar or sheet stock. PTFE is widely used for nonstick coatings, wear pads, and valve and pump seats and seals.

Another member of the fluoropolymer family is fluorinated ethylene propylene (FEP), which is a copolymer of PTFE and HFP (hexafluoropropylene). Its properties are generally inferior to PTFE but it has a distinct advantage in that it is melt processable. This allows it to be used in a number of applications, including tubing, gaskets, pipe liners for chemical processing, and shrink tubing.

Perfluoroalkoxy (PFA) is another copolymer that is melt processable and also bonds well to metals. It has the nonstick characteristics typical of fluoropolymers and is a favored material for nonstick coatings for cookware and industrial applications (PTFE can only be bonded to metal by mechanical means).

The last fluoropolymer we will discuss is polyvinyl difluoride, or PVDF. This material has two fluorine molecules, both on the same carbon atom. Vinylidene fluoride, the precursor to PVDF, is obtained by the hydrofluorination of vinylidene chloride. This material has outstanding flame resistance, having an ASTM E-84 25/50 rating for flame spread and smoke generation and exceptional chemical resistance and impact strength. It is easily melt processable by conventional means. PVDF is used in chemical process piping systems and liners, plenums, containers for chemical storage, batteries, shrink tubing and coatings.

### 3.5.8 High Performance Polymers

The high performance polymers are the high end of the engineering plastics category and display some exceptional properties compared to the majority of plastics materials, usually at a very premium cost. The definition of exceptional continues to change as new and improved polymers are continually being developed. These polymers typically have backbones that contain benzene rings or substituted benzene rings that yield high temperature and mechanical properties. Because of their superior properties they are used frequently as the base resins in many structural composites using long glass and carbon fibers or high aspect fillers.

### *3.5.8.1 Polyphenylenes*

The first family we will discuss is the polyphenylenes which have two principal materials, polyphenylene ether (PPE) and polyphenylene sulfide (PPS). Typical phenylene bonding is comprised of benzene rings joined by one or a combination of oxygens and sulphurs. A schematic of PPS is shown in Figure 3.20.

PPS is a semicrystalline polymer that packs tightly into a crystal lattice. It has very low water absorption, is thermally and dimensionally stable, flame retardant, with outstanding hardness and low shrinkage. Unfilled PPS has a HDT at 264psi of 219 °F (104 °C) and filled versions can be as high as 460 °F (238 °C). It can be used as a metal replacement in many applications but PPS must be processed precisely to develop the crystallinity required to fully achieve properties. A properly molded part when dropped on the floor or a table will exhibit a tinny sound typically associated with metals. Blending PPE with PS yields a material with improved moldability but generally lower physical properties. PPS is used for direct metal replacements, pump impellers and housings, and electrical components. The more easily processed PPE blend is used for computer and business machine housings, video displays and automobile wheel covers (material is plateable).

### *3.5.8.2 Polysulfones*

The polysulfone family also has two principal members, polysulfone (PSU) and polyethersulfone (PES). $SO_2$ group with one or more aromatics creates a stiff strong polymer with a high melting point. A schematic of PSU is shown in Figure 3.21.

The polysulfones are amorphous and transparent with a slight yellow haze. They are more dimensionally stable and tougher than polyphenylenes. They also are sterilizable with good chemical resistance, and are

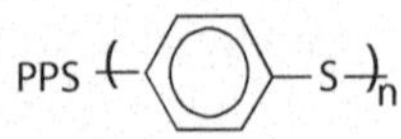


**Figure 3.20** Polyphenylene sulfide.

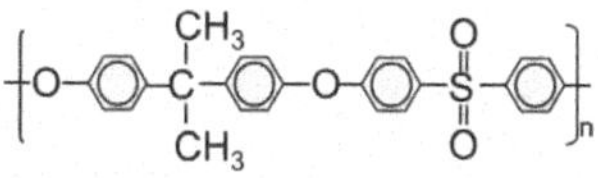

**Figure 3.21** Polysulfone.

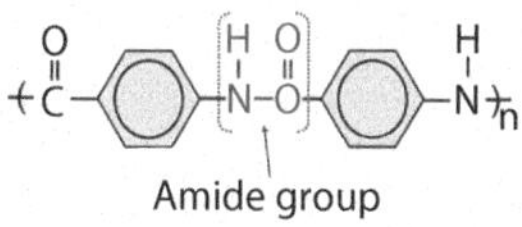


**Figure 3.22** Polyaramide.

FDA approved. PES has a little higher tensile strength and modulus and a little better HDT. PSU has a little better impact strength.

Applications include process pipe in dairies, appliance housing, view ports, dishwasher components and circuit breakers

### 3.5.8.3 *Polyaramids*

Polyaramides are a polyamide polymer with benzene rings between the amide groups (Figure 3.22).

The aromatic groups present in the polymer backbone greatly increase stiffness and toughness, solvent resistance and melting point over a normal polyamide (nylon). Aramids are solvent resistant, have a high melting point and are non-burning.

Polyaramids are used to make very strong fibers that are used for tire reinforcement, suits for race car drivers, fire protective outer wear trademarked as Nomex ™ and bullet proof vests trademarked as Kevlar ™.

### 3.5.8.4 *Polyarylether Ketones*

Polyarylether ketones are polymers made up of benzene rings joined by ether and ketone links and include polyether ketone (PEK) (Figure 3.23) and polyether ether ketone (PEEK).

The materials exhibit excellent stiffness and high modulus and have a higher use temperature than phenylenes. Unfilled PEEK has an HDT @ 264 psi of 306 °F (152 °C) and unfilled PEK has an HDT of 325 °F (163 °C). There also are FDA approved grades.

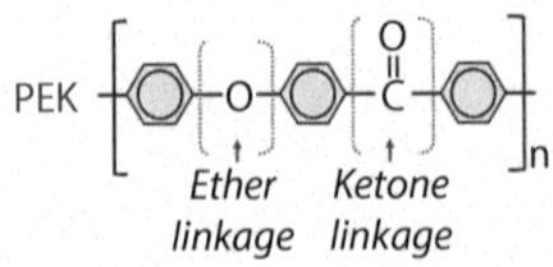


**Figure 3.23** Polyether ketone.

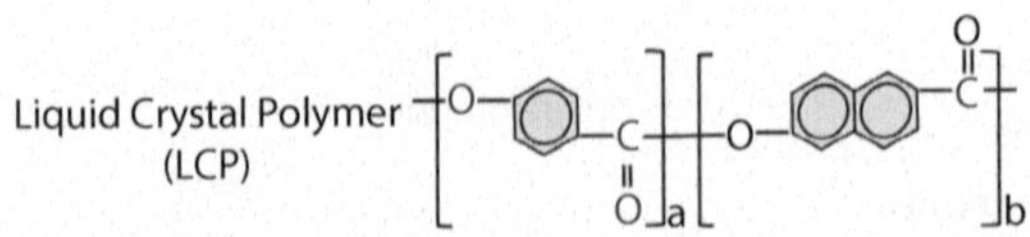


**Figure 3.24** Liquid Crystal Polymer (LCP).

Applications include carbon fiber composites for the aerospace industry, thermal inkjet printer parts, high temperature gears and bearings and coatings.

### *3.5.8.5 Liquid Crystal Polymers (LCPs)*

Although technically a thermoplastic polyester, liquid crystal polymers, commonly known as LCPs, are wholly aromatic co-polyesters that exhibit a highly ordered structure (Figure 3.24) in both the solid state and also as a melt. They range from high melting sinterable products to those that can be melt processed on conventional equipment.

There are several different products available in this family, with the most common based on terephthalic acid, dihydroxybiphenol, and p-hydroxybenzoic acid or p-hydroxybenzoic acid and hydroxynapthanoic acid [10].

LCPs are highly crystalline, inherently flame retardant, thermotropic (melt-orienting) thermoplastics. LCP molecules remain well-ordered even in the melt and slide past each other easily under shear. The result is very low melt viscosity, making it easy to fill extremely thin walls and to replicate intricate features. They exhibit little or no shrinkage in the flow direction and require very little time to set up or solidify. They accept fillers readily and are available in mineral, glass fiber and carbon fiber filled grades.

LCPs have outstanding mechanical properties at high temperatures with a tensile modulus from over 1 million to over 5 million depending upon the grade. At elevated temperatures they retain a modulus of over

500,000 psi at 575 °F (302 °C) and have an HDT @ 264 psi from 450 °F to 560 °F (232–293 °C). They also have excellent electrical properties, chemical resistance, inherent flame retardance (UL94 V0 rating), good weatherability, and resist stress cracking even at elevated temperatures. They do exhibit poor weld line strength, which must be considered in the design process, and the material must be dried prior to processing.

LCPs are used in applications that take advantage of their high temperature resistance and good electrical and chemical properties such as sockets, bobbins, switches, connectors, chip carriers, sterilizable trays, dental tools, surgical instruments and pumps. They are a good choice for potential metal and ceramic substitutions.

### 3.5.8.6 *Thermoplastic Polyimides*

Thermoplastic polyimides (PI) (Figure 3.25) are a family of exceptional high temperature performance polymers. The imide group adds stiffness and restricts molecular movement, causing very high melting points and use temperatures. If the imide group is attached to the chain by an amide group we have a polyamideimide (PAI).

Polyimides have the highest use temperature of any *conventional* thermoplastic and possess a very low coefficient of friction. They are very difficult to process with process temperatures above 700 °F (370 °C). Applications include high temperature, gears, bearings, coatings, sliding parts, sterilizable parts and parts that require extreme dimensional stability.

### 3.5.8.7 *Polybenzimidazole*

Polybenzimidazole (PBI) (Figure 3.26) is a unique amorphous thermoplastic with the highest $T_G$ (427 °C) and highest HDT (435 °C @ 264 psi) of any thermoplastic (*today*)[11].

PBI was commercialized in 1983 by Celanese Corp. as a flame-resistant fiber for astronaut garments (Trade name Celazole™). It is

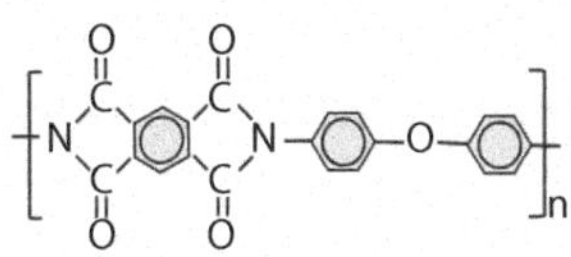

**Figure 3.25** Polyimide.

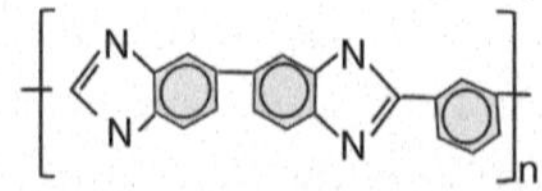


**Figure 3.26** Polybenzimidazole.

extremely difficult to process, with no true melting point. It is processed by compression molding at very high temperatures and pressures or by solution casting. Applications include high performance fibers, bushings and seals, and high purity semiconductor applications.

It can be blended with PEEK which will lower the physical properties, but will improve processability.

## 3.6 Thermoplastic Elastomers

An interesting family of thermoplastic materials is thermoplastic elastomers.

Although they melt and flow like a thermoplastic, because of their exceptional elongation, they are called elastomers. All materials exhibit some elongation, but typically quite small, in the range of 2% or less. Elastomeric materials can be repeatedly stretched over twice their normal length.

Thermoplastic elastomers (TPEs) are a class of polymers that, within their design limits, behave like a thermoset rubber, but above their melt temperatures, are melt processable by standard thermoplastic processing methods such as injection molding or extrusion. Unlike a thermoset rubber, TPEs can be easily reprocessed and remolded. The ability to process these materials as a conventional thermoplastic allows for design and fabrication freedom that thermoset rubbers do not offer.

In thermoplastic elastomers, there is the development of domains of hard and soft regions at the microstructural level. When heated, these domains are miscible, but separate upon cooling. Each of these polymers is a block copolymer with both hard and soft segments. The soft segments are usually aliphatic and linear. The hard segments are usually aromatic with polar groups that will form hydrogen bonds.

There are several different types of thermoplastic elastomers which can be used in many applications as a rubber substitute. Styrene-based elastomers, such as Kraton™ (trademark of Kraton Polymers U.S. LLC), is inexpensive and used in low cost footwear and a variety of molded and extruded parts. Urethane-based elastomers, such as spandex and Lycra ™ (trademark of Invista North America), have excellent abrasion resistance, higher use temperature and fatigue resistance. Polyester-based elastomers, such as Hytrel® (trade mark of E. I. Du Pont de Nemours & Co.), have good stiffness and resistance to some chemicals and are used in flexible tubing and fittings to handle chemicals. A recent entry to TPEs are olefin block copolymers (Dow INFUSE®). They have chains with alternating blocks of "hard" (highly rigid) and "soft" (highly elastomeric) segments. They offer good compression set and elastic recovery properties and are easy to process.

## 3.7 Biopolymers

There is growing interest and industry capability in the area of "Green" or renewable plastics, although there can be confusion about what this means. Most plastics are based on feedstocks based on petroleum or natural gas. There is considerable interest and activity into moving away from these and in using a renewable feedstock source such as corn or soybeans for plastic materials. Materials made in this manner are called biobased.

Similarly, there are justifiable concerns that plastic materials that enter the waste stream do not degrade over time and can cause environmental damage. To address this, material suppliers are developing polymers that will break down into basic components that will not harm the environment. These polymers are called biodegradeable. The ASTM (American Society for Testing International) is developing test methods to define and quantify this attribute.

The next three materials discussed are commercial biobased polymers that are available today. There are others (nylon 6/11 was already discussed in the section on nylons) and the list is growing. Also, a number of manufacturers are introducing partially biobased materials, where one or more of their components are made from a renewable resource.

### 3.7.1 Polylactic Acid

Polylactic acid (PLA) is a polyester biopolymer that is made up of long molecular chains of the polymer polylactide. PLA is derived from naturally occurring plant sugar [12]. It is currently the most commercially important biopolymer, although significant developments in this area are rapidly expanding the choices available.

PLA has a HDT @ 264 psi of 50 °C (122 °F) and a notched Izod impact of 0.3 ft lbs/inch [13]. PLA will biodegrade to carbon dioxide and water, but it does so very slowly.

Applications include fibers for textiles, film and packaging applications, bottles, small moldings and filaments for 3D printing. It can be used as an alloying resin with other plastics to achieve a partially biobased material, for example, Arkema's Plexiglas® Rnew®, which is an alloy of PLA and acrylic.

### 3.7.2 Polyhydroxyalkanoates

Polyhydroxyalkanoates (PHAs) are a family of polyhydroxyesters that are made by fermentation using renewable carbon-based feedstocks, such as methane, making them 100% biobased. They also are biodegradable in soil and marine environments.

PHA has HDT @ 264 psi up to 170 °F and a notched Izod impact of 0.49 ft lbs/inch and a tensile strength of 3625 psi [14].

Applications include food service and packaging, including caps and closures, and disposable items such as forks, knives, tubs, trays, jars, packaging film, fibers, utensils, 3D filament, and biodegradable agricultural items.

### 3.7.3 Polybutylene Succinate

Polybutylene succinate (PBS) is a new entry in the growing base of biopolymers. It is produced by the polymerization of succinic acid and 1,4 butanediol. A member of the polyester family, it is biodegradable, decomposing into carbon dioxide and water. Initially, the PBS monomers came from petroleum, but are now available with biobased succinic acid and an all-biobased material is planned shortly.

PBS properties are similar to polypropylene. The material accepts fillers well and is used as a blending resin with PLA. It is used in applications such as agricultural sheet and film and automotive molding compounds.

## 3.8 Thermosets

Polymeric resins that are capable of forming crosslinks are called thermosets. The process of crosslinking is called curing. Thermosets, when heated, may soften slightly but will retain their shape until decomposition. It is the chemical reaction that creates the crosslinks between the polymer chains that cause these materials to harden and the process is irreversible. Curing agents (initiators) are commonly used to start the crosslinking process.

When thermoset polymers react, usually heat is given off as part of the process. This is called *heat of reaction* and the heat given off is called an *exotherm* (exothermic reaction). Thermoset resins can be used in several different forms or stages. In an "A Stage resin" the material is thin (fluid) and is largely monomer. In a "B Stage" resin, the material is partially polymerized. This is also called "B-Staging." Resins for a molding process are "B Stage" resins. A "C Stage" resin is one where the material is fully cured.

### 3.8.1 Phenolics

Phenolics (PFs) were first synthesized by Dr. Leo Baekeland in 1907 (Bakelite™). They were the first plastic to be produced from synthetic materials. Phenolics are formed by the condensation polymerization of phenol ($C_6H_5OH$) and formaldehyde ($CH_2O$).

Phenolics were fairly inexpensive and were relatively easy to compress and transfer mold. As a result, throughout the early 1900s, phenolics were the highest volume plastic material. They were used in telephones, radios, electrical insulators and early 78 rpm records. Newer engineering thermoplastics have replaced phenolics in many applications but because of their excellent heat resistance, flame resistance, electrical insulation properties and adhesive characteristics they still are the material of choice for a large number of applications. One of the disadvantages of phenolics is that they are limited to generally darker colors. Phenolics are used in molded disc brake cylinders, saucepan handles, electrical plugs and switches, electrical iron parts and ablation shields. They also are used in a number of coating and paper impregnation applications and lamination applications, including plywood. Phenolics can be mechanically whipped into foam and cured. The foam product will absorb and hold water and is widely used as floral foam. Figure 3.27 shows several applications.

### 3.8.2 Amino Plastics

Amino plastics are formed by the condensation polymerization of multifunctional monomers containing amine groups ($NH_2$) and

**Figure 3.27** Phenolic (PF) products.

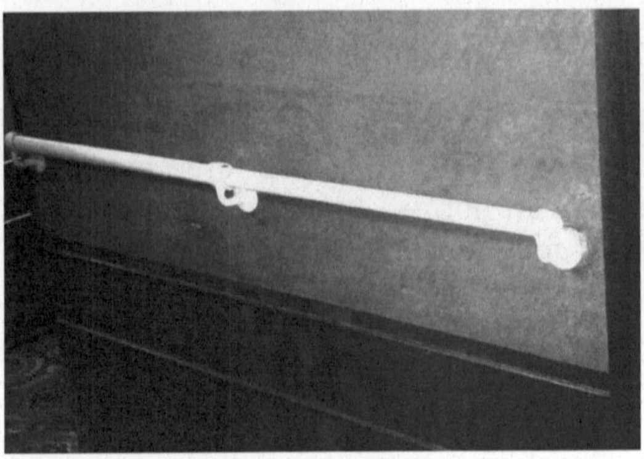

**Figure 3.28** UF plastic handrail on the *Queen Mary*.

formaldehyde. Amino plastics are lighter in color than phenolics, which is important in decorative applications, and they process very similarly to phenolics. They have slightly better arc resistance and adhesive properties than phenolics, but are a little more expensive.

The two amino plastics of interest are urea formaldehyde (UF) and melamine formaldehyde (MF).

Patents were issued to German and British chemists for urea formaldehyde in the early 1920s and UF molding resins were first introduced in the mid 1920s. This provided a plastic molding resin that was available in lighter colors. Figure 3.28 shows a plastic handrail made of UF on the cruise ship *Queen Mary*, built in the early 1930s.

Urea formaldehyde is less expensive than MF, slightly more than phenolic. It is used for bottle caps, insulation and adhesives.

**Figure 3.29** Amino plastics (UF and MF).

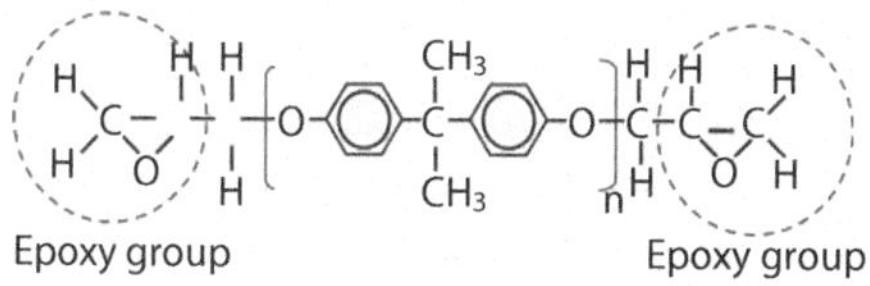


**Figure 3.30** Three-member epoxy ring on each end of polymer chain.

Melamine formaldehyde (MF) is also called melamine. It has outstanding surface hardness, excellent colorability and adhesive characteristics. This has led to its widespread use as a countertop material (Formica®, trademark of Formica Corporation). It also is used for molded dinnerware (Melmac®, trademark of American Cyanamid Company) and automotive coatings. Figure 3.29 shows several applications of UF and MF materials.

### 3.8.3 Epoxies

Epoxies are characterized by the presence of a unique three-member epoxy ring on each end of the polymer chain (see Figure 3.30).

Crosslinking of this thermoset is initiated by a material with a reactive group called a "hardener" which opens up the epoxy ring. The epoxy resin and hardener are sometimes called "part A" and "part B." To achieve optimum properties, balance between epoxy sites and hardener sites is critical for proper curing and maximum property development. Epoxy can also be combined with reinforcements for molding. This is called a prepreg.

Epoxies are widely used as high quality industrial grade adhesives. Other applications include liners for containers, high quality paints and coatings, and structural components for aerospace. Because of their low viscosities and easy flow and low temperature cure, epoxies are used for encapsulation and potting of electronic components as well as casting applications (see other prototyping methods in Chapter 4).

### 3.8.4 Thermoset Polyesters

Thermoset polyesters or unsaturated polyesters (UP), like thermoplastic polyesters, are formed by condensation polymerization. With thermoset polyesters, the polymerization reaction is terminated to produce a short-chain, low-molecular-weight liquid polymer. This polymer is able to crosslink because it contains an active carbon-carbon double bond, as shown in Figure 3.31. A double bond is said to be an "unsaturated" bond, so these materials are called "unsaturated polyesters."

The double bond will react by addition polymerization with the addition of an initiator, usually a peroxide. This is important because no condensation byproduct is formed. When an unsaturated polyester resin is cured, there is no water or other byproducts that need to be removed which could cause bubbles in the finished product. When these unsaturated polyesters are mixed with fiber reinforcements the resultant product is called a *composite,* or when glass is used, a fiberglass reinforced plastic (FRP). These products are used to make glass-wrapped pipe and fittings, boats, and auto bodies.

There are two basic kinds of UP, alkyds and allylics. Alkyds can be made into molding compounds by blending resin, initiator, fillers, pigments, etc., which are mixed on a heated roll, cooled and ground to a fine powder. They have high electrical resistance, low moisture sensitivity and good dimensional stability. They are also used for high durability paints. Allylics can be compounded to a paste form for body putty or used in liquid form for potting and encapsulation.

UPs can be mixed with fillers and long fiber reinforcements to make a bulk molding compound (BMC), also called dough molding

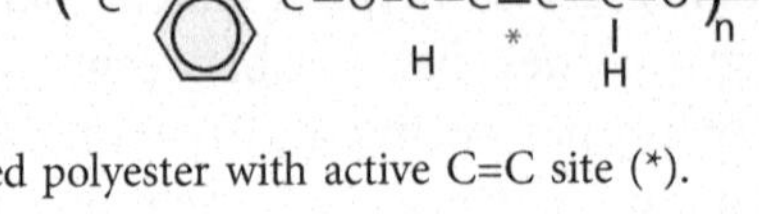

**Figure 3.31** Unsaturated polyester with active C=C site (*).

compound or premix, which can be compression molded. A sandwich of resin and chopped glass can also be formed into continuous sheets of molding material layered between polyethylene sheets, called a sheet molding compound (SMC). The sheets are then cut to length, the PE sheets removed and the sheet of resin and glass is placed in a compression mold and formed and cured into its final shape. This is the process that is used to make the Corvette body.

Vinyl esters (VEs) are also in the polyester family but have some distinct advantages over standard UPs. They are slightly more expensive but have superior toughness. They have outstanding chemical resistance and are a material of choice for corrosion resistant applications like pipe (FRP pipe), fittings, chemical storage containers and reaction vessels.

### 3.8.5 Thermoset Polyurethanes

Thermoset polyurethanes are produced through the reaction of a multifunctional alcohol (polyol) plus an isocyanate. They combine similarly to condensation polymerization, except no water is given off, which is an advantage when making a finished product. Flexibility of urethane chemistry allows tailoring a wide range of properties to fit a diverse variety of applications. Isocyanates are reactive and toxic and must be handled with appropriate precautions.

The liquid and reactive nature of the two urethane components can provide some excellent process opportunities. These properties make PU particularly suited to reaction injection molding (RIM), a low pressure, fast process. This process can be used for making fairly large molded parts such as auto bumpers. The material also works well in casting applications.

By incorporating a chemical blowing agent PU can be made into a number of foamed products. Flexible foams with densities from 1 to 12 lbs/cu ft can be used for bedding, packaging, vehicle seating, furniture, insulation and carpet padding. Rigid foams are used for construction, insulation, packaging and transportation.

Other applications include thermoset urethane elastomers, a rubber substitute, that can be made into rollers, wheels, bumpers, and athletic shoe components, similar to thermoplastic urethane elastomers (TPU). PU also can be used to make fibers such as "spandex" and tough and abrasion resistant coatings. UV sensitivity will limit possible outdoor applications.

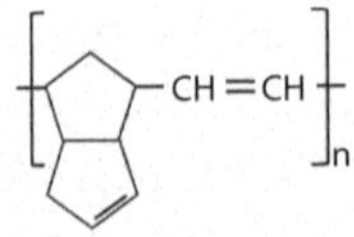

**Figure 3.32** Polydicyclopentadiene.

### 3.8.6 Polydicyclopentadiene

Polydicyclopentadiene (DCPD) (Figure 3.32) is a relatively new material that is formed through ring opening metathesis polymerization (ROMP).

DCPD is processed by reaction injection molding, which allows it to be used to make very large parts. The material possesses excellent impact strength down to −40 °C, excellent stiffness, low density and good chemical resistance to acids and bases [15].

Applications include bodies for large tractors, construction equipment, trucks and storage tanks.

### 3.8.7 Thermoset Polyimides

Similar to their thermoplastic cousins, thermoset polyimides are polymers for extreme high temperature applications (>600 °F or 315 °C). They can be crosslinked by both condensation and addition mechanisms. Addition polymers are easier to use but have slightly lower properties.

Applications include aircraft engine exhaust ducts, high speed aircraft skin panels and high temperature electrical components.

## 3.9 Fillers and Reinforcements

While fillers and reinforcements could be a topic for a chapter (or a book) unto themselves, the purpose here is to make the reader aware of the virtually infinite possibilities for cost improvement and property enhancement that can be achieved by compounding plastic materials with fillers and reinforcements.

A filler can be defined as an inert material that does not have any substantial reinforcing capability. They are typically particles with no specific length-to-width ratio, unlike a fiber. Fillers are generally added to reduce costs, although they can impart property enhancements such as flame retardance, color, opacity, heat stability or UV resistance.

Common fillers used for cost reduction are clay, calcium carbonate (ground limestone), talc, glass beads and wood flour. A common example would be the use of calcium carbonate in PVC water pipe and electrical conduit.

Flammability is often a concern when designing a plastic part. Flammability can be improved by adding organic halogenated compounds, usually containing bromine or chlorine, which slow the oxidation reaction, aluminum hydroxide and magnesium hydroxide which remove heat as they break down, or organophosphate intumescent additives which separate the fuel from the fire. Plastic compounders offer a wide array of flame retardant compounds for many different applications.

Color and opacity is obtained with a wide range of chemicals from titanium dioxide for white to carbon black for black and the complete range of colors in between. Pearlescent and fluorescent effects can be obtained in a variety of polymers. There are also materials available that are thermochromic. These materials will change color with changes in temperature and can be used for a number of interesting applications such as safety sensors, temperature limit sensors and thermometers.

Heat stabilizers are used to prevent damage to the polymer chain caused by the generation of free radicals during processing. Compounds typically used include organotins and antimonies.

Plastics are used in applications such as bearings and gears to take advantage of their natural lubricity and low coefficient of friction. These properties can be further enhanced by adding fillers such as PTFE, graphite powder and silicon. See Chapter 11, Bearings, for more on this subject.

UV stability can be improved by the addition of a UV absorber, such as carbon black (preferentially absorb UV), or by adding a hindered amine light stabilizer (HALS), which inhibit UV effects by scavenging and quenching free electrons.

Another very useful application of fillers is to improve the dimensional stability and reduce the shrinkage of a part. High loading of low aspect fillers like glass beads can significantly reduce the amount of shrinkage in a plastic. When used in high shrink materials like polypropylene, they can enable molding parts that would otherwise be impossible.

Fillers are mixed into the plastic in its liquid state, either by a compounder or directly by the processor.

Impact modifiers are materials added to a plastic to increase its toughness by improving the plastic's ability to absorb impact energy. The most common materials used are rubbers that are added in small particulate form and mixed into the polymer melt as a filler. They can also be incorporated by copolymerization of a plastic, as in acrylonitrile butadiene styrene

(ABS) where butadiene rubber is copolymerized onto a polystyrene backbone.

A plastic material's strength and modulus can be substantially improved by adding strong high modulus materials called reinforcements. Reinforcements usually are in the shape of fibers and they can impart very high strength (in one direction). Common fibers that are used include glass, carbon and metals. If the fiber content is high, the fiber properties will dominate, if it is low, the plastic properties will dominate. Plastics with reinforcements are sometimes called *composites.* Glass fibers have been the workhorse due to the relatively low cost and the significant property improvement they impart. An unfilled nylon 6/6 compound will have a tensile strength of approximately 6000 psi. Compounding with 30% glass fibers will increase that to 25,000 psi. Carbon fibers once used sparingly because of cost are now finding extensive use because of their ability to deliver premium strength to weight properties when compounded with engineering resins and thermosets. Carbon fiber composites are playing a key role in helping auto manufacturers meet the latest Corporate Average Fuel Economy (CAFÉ) standards and they are being widely used in structural components for aircraft.

Like fillers, reinforcements are incorporated into the plastic in its liquid state, either by a compounder or the processor. Care has to be taken to minimize fiber breakage in this process, which would result in a reduction of physical properties.

Unlike an unfilled plastic whose properties are generally isotropic (the same in all directions), when using a filled plastic it is important to recognize that the material properties could be orthotropic, different in two or more directions. This is especially true with fiber-filled materials. Usually, the higher the concentration and the longer the fibers are, the greater the difference in properties in the direction of fiber orientation (flow direction) from the perpendicular to the fiber orientation (across the flow direction). In extreme cases of fiber loading, these properties can differ by a factor of ten or more. This behavior can also be affected by the process selected and the process conditions within a particular process. Adequate analysis of these types of materials generally requires a computer-aided finite element analysis. To ameliorate this problem, woven fiber cloth is sometimes used when making a composite part.

# 4

# Process Overviews, Advantages and Constraints

One of the biggest advantages that plastics materials give to the design engineer is that they are relatively easy to inexpensively form into a usable shape relative to many other competing materials. The purpose of this chapter is to give an overview of the primary processes available to make plastics products with and to discuss advantages and limitations of each process from a design perspective.

There are a number of manufacturing processes that are commonly used as well as several special purpose processes that provide manufacturing solutions for an almost infinite range of shapes, sizes and materials. What most of them have in common is that they soften or melt the plastic material and form it into the desired shape. What many of these processes provide is a means to directly produce a net shape part, one that requires no further operations for it to be usable.

We can divide the processes we will discuss into two basic types, continuous and cyclical. A continuous process is one that, once started, produces a continuous stream of product that will be similar over time.

A cyclical process is one that produces a series of discrete individual products. The process basically starts over every time a part or group of parts is made. This will be of importance when discussing individual part quality.

## 4.1 Extrusion

The first process we will discuss is extrusion. Extrusion is a continuous process that will deliver a continuous stream of a polymer melt that can be subsequently formed into a desired shape. A good analogy would be squeezing toothpaste from a tube. As long as you continue to squeeze and you still have toothpaste in the tube, a stream of toothpaste will come out the opening at the end. Figure 4.1 gives a simplified depiction of a typical extrusion line.

Plastic material is brought into the feed hopper where a rotating screw will convey the material forward through the barrel where it will be melted by a combination of shear, frictional and conductive heating. It will then be forced through the extrusion die that will give the melted material its desired shape. As the shape leave the die, it will be cooled, possibly with a water bath, a cooling table or a stream of air. The puller after the cooling step is a mechanical device that will grip the profile and pull it continuously from the upstream process and push it to the downstream process. Downstream from the puller could be cutoff equipment, coiling or packaging equipment, and/or a variety of secondary processes, including machining or decorating. Figure 4.2 shows a picture of an actual screw and barrel

Extruded shapes are generically called profiles and could include tubes, bars, sheets or any continuous shape. The extrusion process is a widely

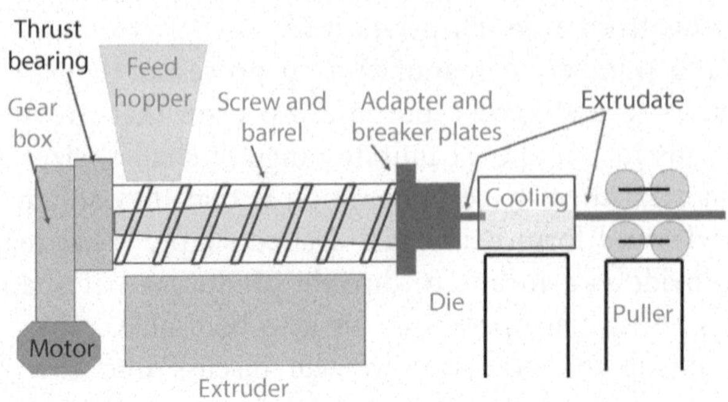


**Figure 4.1** Extrusion line.

**Figure 4.2** Extrusion screw and barrel.

used manufacturing process. The material selection can be any material that can be made to flow, including a wide selection of plastics, metals and liquids. Most thermoplastics and some thermosets can be practically extruded. A high molecular weight (low melt index) is desirable for best extrusion, giving good melt strength, allowing the material to hold shape as it leaves the die. A broad molecular weight distribution helps facilitate the melting of high molecular weight materials and provides lubrication of the process as well as also helping the molten material to hold its shape. Some very heat-sensitive materials (PVC, CPVC) are usually extruded by a twin screw extrusion process, which heats the material more gently. Some difficult, hard to flow materials (PTFE, Polyimide) are extruded by a ram extrusion (as opposed to screw extrusion) process, where the material is advanced down the barrel by a hydraulic ram instead of a rotating screw.

The extrusion process allows making a continuous part that can be solid (a solid bar or a sheet), hollow (a pipe or tube) or a profile (siding, window profiles, I-beams). Although somewhat limited in the forms it can make, extrusion can be an extremely low cost method of manufacturing, particularly on higher volume production runs. Anything other than a continuous profile or shape, such as a corrugated tube or drainage pipe with cross holes, will require additional tooling or equipment and an operation downstream from the extrusion die.

From a product design standpoint, a constant material thickness through the part cross section is most desirable. A partially closed section is easier to produce than a fully closed section and true sharp corners cannot be made. The surface finish that can be achievable is a function of the polish on the extrusion die and the sizing/cooling device. If a textured surface is desired it can be added in a downstream operation with embossing rolls.

Besides profiles and tubes, there are several other extrusion process variations of interest. The first is blown film. Instead of extruding a thin sheet to make a film or bag, a thin tube is extruded. As it is extruded (usually vertically) the tube is pinched closed by nip rollers to form a bubble that is stretched and expanded by internal air pressure. The bubble is supported by a tower structure, which provides sufficient time for the plastic film to cool before it is pinched together again at the top of the bubble. This allows the manufacture of a very thin film. This is the method used to make thin-wall high volume films used in garbage bags and grocery bags. Figure 4.3 shows a blown film line. You can see a film bubble exiting the die as well as the cooling tower, which in this case extends vertically about thirty feet above the extrusion die.

Extrusion is also widely used for coating processes, including wire coating, paper coating and fabric coating. In these processes, the substrate (wire, paper or fabric) is passed through a die called a crosshead die and the extruder will apply melted polymer to it. Figure 4.4 shows a schematic of a wire coating process.

Products made up of two or more materials can be manufactured by using a co-extrusion process. In this process we feed the extrusion die from two or more extruders. The melt streams can be combined upstream of the die through a special adapter or entered directly into a crosshead die. A tubular crosshead die is used for multilayer blown

**Figure 4.3** Blown film tower.

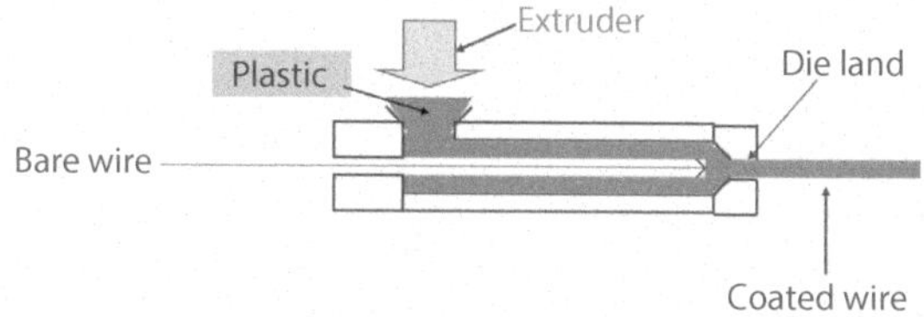


**Figure 4.4** Wire coating schematic.

**Figure 4.5** Co-extrusion.

film. We can also use this process to make a foam core product, add an internal layer of regrind (reprocessed material), seals and gaskets with a hard backing and soft seal, and for cosmetic marking or striping. Figure 4.5 shows a co-extrusion operation making pipe with a colored stripe to identify the intended application.

In addition to being a primary process onto itself (called extrusion molding or, more commonly, extrusion), an extruder is a key component of many other primary processes, including injection molding, blow molding, injection blow molding and foam molding. Also, extrusion is a very efficient mixing process and is used to compound plastic materials with a wide variety of additives and colorants.

## 4.2 Injection Molding

Injection molding is a cyclical, discrete process. The molding process itself consists of melting the plastic in an extruder much like the one we discussed previously, forcing the melted plastic into a mold, solidifying the part in the mold by cooling (a thermoplastic) or heating (a thermoset), opening the mold and removing the part or parts, closing the mold and repeating the cycle again.

Figure 4.6 gives a simplified depiction of a typical injection molding machine. As in the extruder, the material is brought into the machine through the feed hopper where it is delivered to the screw and barrel. The screw and barrel prepare the melt and deliver it to the mold which is mounted between the fixed and moving platen. As the moving platen is moved to the right by a hydraulic cylinder (or a ball screw in an all-electric machine) the mold is opened to remove the part and then moved to the left to reclose the mold in preparation to make the next part.

In the injection molding machine the screw and barrel function a little differently than in an extruder. The melt is achieved by the friction between the rotating screw and barrel and the external heat that is applied to the barrel through electrical resistance heaters, as in the extruder, but as the material is conveyed forward by the rotating screw, the screw is forced back in the barrel to make space to accommodate the melt. When adequate melt is prepared for the part to be made, the screw rotation is stopped and the screw is brought forward to push the melt into the mold.

If you look at Figure 4.7, in step 1 the screw is fully withdrawn and a pool of melted plastic is in the front of the screw (shown in red). In step 2, the screw is pushed forward, forcing the melted plastic into the mold. If the part being made is thick, at this point the screw may be held forward under pressure for a time to make sure the gate area (the entrance area into the part) has cooled enough to prevent backflow of the melt. This is known as *holding time.* In step 3, the

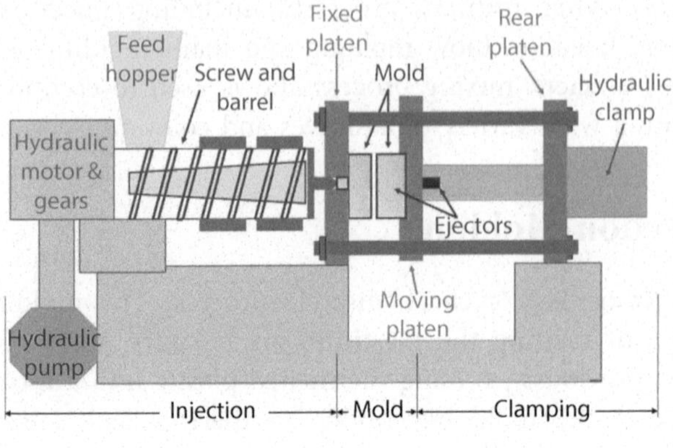


**Figure 4.6** Injection molding machine.

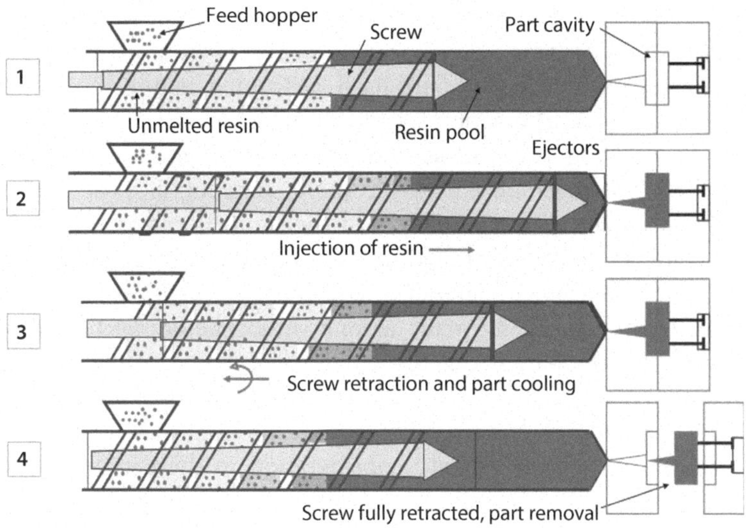


**Figure 4.7** Reciprocating injection process.

screw is rotated as a new melt pool is being prepared for the next cycle. While this is happening, the mold remains closed as the melted plastic is being cooled by water (usually) circulating through the mold to harden the part. In step 4, the screw is fully retracted and a melt pool is available to make the next part. At this point the mold is pulled open, the cooled part removed, the mold is reclosed and we return to step 1 to restart the process. Because of the back and forth action of the screw this is called a reciprocating screw process and the machine a reciprocating screw injection molding machine. As the process is restarted after every cycle, it is important to have good controls on all aspects of the machine function to insure a consistent melt quality and a consistent part.

If a thermosetting material was being used, the mold would be heated with electric heaters or hot oil to crosslink and cure the thermoset part in the mold. The part would be ejected hot from the mold.

A designer should be aware of a few terms that are used in injection molding. The *shot size* of a machine is the amount of plastic in ounces or grams that a molding machine can produce in one cycle. Ideally a part plus the sprue and runner weight will be between 25% and 75% of the maximum machine capability. The cycle time is the time it takes to move the resin into the mold, hold the material under pressure until gate seal, recover the material for the next cycle and open and close the mold (steps 1–4 above).

In injection molding, the part itself is formed in the mold or tool. Figure 4.8 shows a schematic of a typical injection mold. At the middle left of the drawing you will see the sprue bushing. This is where the plastic from the machine enters the mold, forming a sprue, shown in red. The sprue could feed directly into the molded part if one part at a time is being made, or, as shown in this drawing, it could feed a runner system which distributes material to each of the individual mold cavities if multiple parts are being made at one time. The connection between the runner and the part is called the gate. The design of the gate geometry is important because it controls how the plastic melt is fed into the part and the gate will leave a mark on the finished part. This gate size and location must be taken into consideration in the design of the finished part. An alternate tool design would combine the sprue and runner into an enclosed heated manifold that keeps runners in molten state through the use of cartridge heaters within the tool. This is called a hot runner system. Hot runner molds are more expensive than cold runner molds and a little more difficult to operate, but they eliminate the need to handle the runner system and can have slightly faster cycle times. From a part design standpoint, there will be either a mark or small residual piece of plastic where the hot runner gate enters the finished part. The mold half that the sprue enters fits against the fixed platen of the molding machine and is called the "fixed" side, also sometimes referred to as the "A" side.

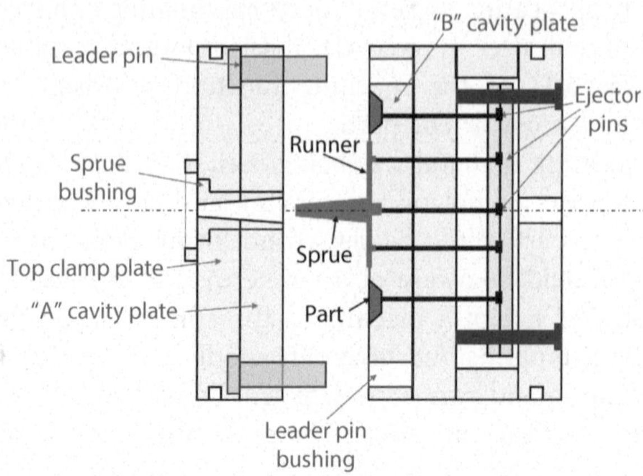


**Figure 4.8** Injection mold.

The gate placement and size, with either a cold or hot runner mold, can affect part quality, part shrinkage and molding cycle. The gate should be located in the thickest section of the part if possible. Sometimes a design will require a compromise between the ideal location for part cosmetics and the ideal location for processability of the part.

Gate location will also affect weld lines. A weld line is the point where two plastic melt streams come together. This will sometimes leave a visible line and will likely be the structurally weakest area of the part. Weld lines will be discussed in more detail in Chapter 5, General Design Considerations.

Most parts will require some kind of mechanical assist to push the parts out of the mold. This assist can be in the form of ejector pins (as shown in Figure 4.8), sleeves, or a number of other means. From a design point of view you want to locate these on non-visual or non-functional surfaces where possible, but again, this sometimes requires a compromise between the ideal location for part cosmetics and the ideal location for the function of the process. Ejector systems will also be discussed in more detail in Chapter 5. The ejector half of the mold usually is located on the moveable platen of the molding machine and is called the "moving" half or also the "B" side of the mold.

When the mold is closed, prior to the injection of plastic, the mold cavity is filled with air. This air has to be removed as the plastic enters the cavity. If the air is trapped, it will compress and heat up, possibly burning the incoming melt, preventing the melt from completely filling out the mold cavity, and possibly preventing any weld lines from reaching optimum strength. To allow the air to escape the cavity, a clearance called a "vent" is machined into the mold. Vents are typically small slots 0.0005" to 0.004" deep on the face of the mold where the cavity halves come together or can be machined on the ejectors or other moving mold parts. Figure 4.9 illustrates this concept.

The injection molding process can be used on a wide variety of materials, including a wide variety of thermoplastics as well as some thermosets. Solids and liquids can be injection molded. The injection molding process is highly repeatable, very precise, yields net shape parts (no secondary operations required) and can be automated. It can be used to produce very complex shapes and parts with variable cross sections. Although very complex parts can be made, the process and part will be optimized if the design can be made so the part has a constant material thickness and avoids very thick sections (use core outs when possible). Thick sections can cause sink marks or dimples due to material shrinkage. Ribs on "T" sections need to be <2/3 of nominal wall (ref Chapter 5). Sharp corners need to be avoided.

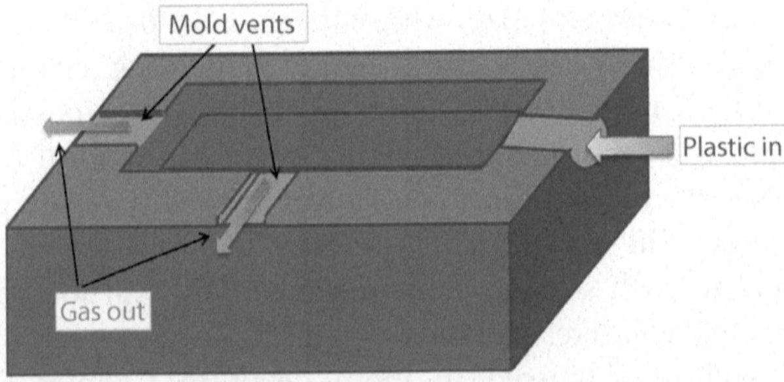


**Figure 4.9** Mold cavity venting.

**Figure 4.10** Co-injection molded part.

Fillets and radii need to be as generous as possible but should be at least 0.020″ (0.5 mm). Large gates are needed to pack out thick parts but will increase cycle time and affect the part appearance. The surface quality achieved will depend primarily on the tool surface and to a lesser extent the material specified.

There are a few modified special injection molding processes that should be mentioned. *Co-injection Molding* is analogous to co-extrusion. Multiple resins are introduced into a single die using 2 or more (injection units). The plastic melts can be mixed at the nozzle, the sprue, or in the mold itself (this is also called "Two Shot Molding" – example: computer keys). The plastic materials need to be compatible or separation between the materials can occur. An inexpensive or recycled material can be used for the core material, a more expensive material used for the outer skin. Also, the core material could be foamed (see foam molding) to reduce cost, add structure, and reduce weight. Figure 4.10 shows a co-injected bottle cap with a blue, softer thermoplastic elastomer to give a good grip to unscrew the cap and a harder plastic underneath to give strength and rigidity.

*High Pressure Foam Molding* creates a foam structure on the inside of a part by either of two ways. In the first, an additive is used that gives off a gas when heated as the plastic is melted (chemical blowing agent or CBA). Although CBAs will create a foam throughout the part, the foaming will tend to be more concentrated in thicker sections. Weight reductions from 3–10% can be realized. The process tends to give a lower stressed part than a solid molding but a poor surface finish will result, as some of the gas bubbles tend to break out on surface.

The second method involves injecting a very high pressure gas in the melt stream at the nozzle. This is a patented process by Trexel (MuCell® microcellular foam). This process will give a greater weight reduction than CBAs, a finer foam structure and a generally better surface appearance.

Using molded foam is a good design tool to reduce weight and molded in stresses in a part and can also be used to achieve cost reduction by minimizing the material in a part design.

*Gas Assisted Injection Molding* is a technique used to reduce sink marks due to shrinkage in thick part sections and is also very effective in reducing the amount of stress in a molded part. In the gas assist process (Figure 4.11), a gas is injected into the melt stream of a partially filled mold cavity. The gas will travel to the lower melt viscosity material in the center of the part, pushing material in the thickest sections against the cavity walls.

Gas can be introduced through the machine nozzle, in the sprue or runner as shown, or directly into the part through a specially designed gas injection port. An inert gas, usually nitrogen, is used to prevent oxidation of the hot plastic material. The process allows the designer to have thicker than normally recommended sections in the part. In addition to reducing sinks and pits on the surface and lowering molding stress, the process will improve tolerance capability.

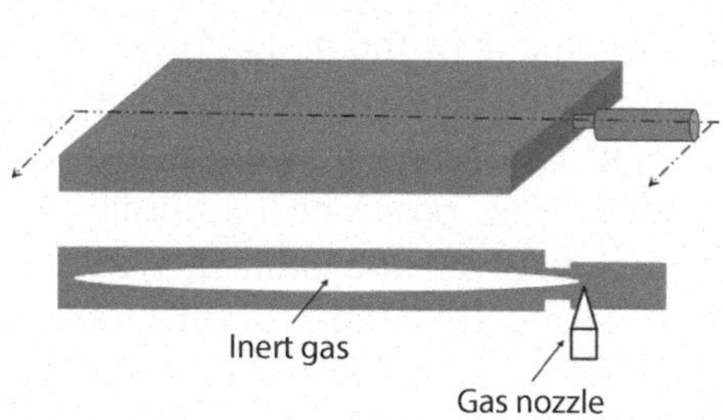


**Figure 4.11** Gas assist injection molding.

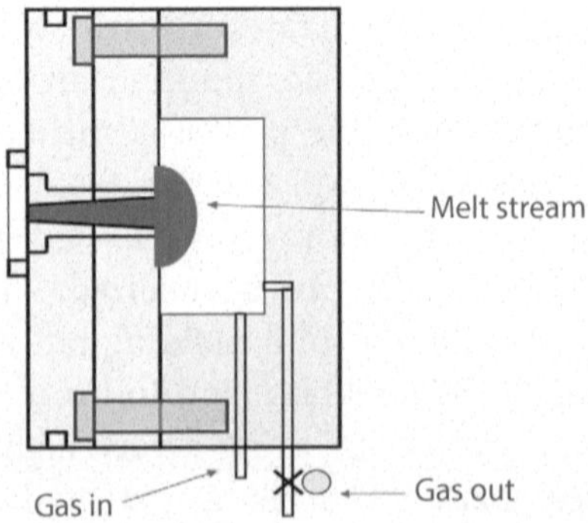


**Figure 4.12** Gas counterpressure.

*Gas Counterpressure Foam Molding* is used to address the problem of a poor surface finish that is typically seen when molding a foamed part. In this process (Figure 4.12) a gas is injected into the mold cavity prior to the injection of the plastic. As the plastic enters the cavity, the gas is allowed to escape as the cavity is filled, maintaining a uniform pressure against the melt. When the cavity is 95–99% filled, the gas pressure is released. The gas pressure prevents gas bubbles from forming at the plastic melt front, which allows an unfoamed skin of plastic melt to form on the cavity walls.

This makes the majority of the foaming take place within the part itself and not on the surface of the part. This allows making a foamed part with a high quality surface finish.

## 4.3 Extrusion Blow Molding

Extrusion blow molding is a cyclical, discrete process that can make simple to medium complex hollow parts. It allows the designer to make one-piece parts with narrow openings and wide, hollow bodies. Figure 4.13 shows a schematic of the basic process.

The basic extrusion blow molding process consists of 3 basic steps. In step 1 the resin is melted in an extruder to form a plastic tube called a "parison." The extruder is positioned vertically and the parison is pulled down by gravity between the mold halves. In step 2, the mold halves close on the parison, pinching off the top and bottom of the tube. In step 3, air is blown into the pinched-off tube through a small hollow tube or blow pin, forcing the tube to expand out to the shape of

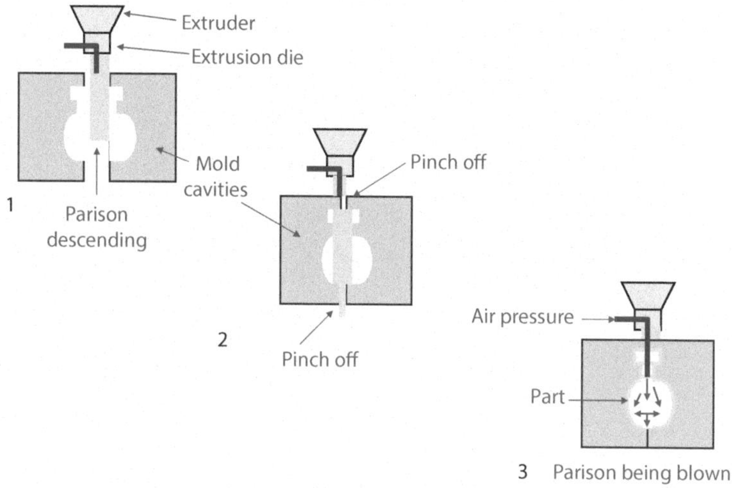


**Figure 4.13** Extrusion blow molding process.

the mold cavity. The mold cools the plastic, opens, and the part is ejected. The steps are repeated to produce additional parts.

There are several versions of the basic process, including the rising mold system where two or more molds are mechanically moved into place to capture the parison and then moved away to complete the blowing process. The multiple mold system is similar to rising mold system, except that many molds are mounted on a rotary table like a wheel.

This process, depicted in Figure 4.14, is very good for producing high volumes.

Most thermoplastics can be blow molded but polymers with good melt strength and a broad MWD work best. Common examples would include PETG soap and detergent bottles, HDPE milk bottles, toys, storage drums, nylon auto coolant bottles and power steering reservoirs. Multiple resins can be used (co-blow molding) where a multi-material parison is extruded to take advantage of multiple material properties (for example, a barrier resin and a high impact resin).

When designing an extrusion blow molded part it should be symmetrical with the wall thickness limited to no more than 1/2″ (12.5 mm). A uniform wall is desired from a design standpoint, but parison sag, particularly on large parts, must be considered. Corners and edges should be generously radiused. A true flat bottom is very difficult to achieve and should be designed to be at least slightly concave. The blow-up

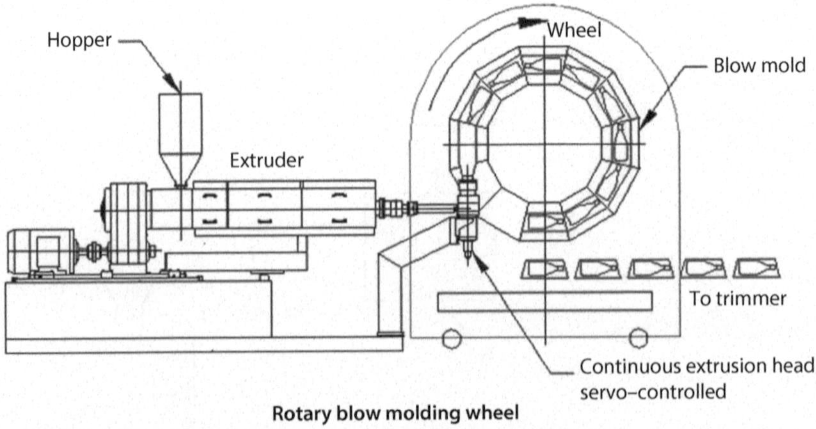


**Figure 4.14** Multiple mold extrusion blow molder.

**Figure 4.15** Examples of blow-molded products.

ratio (mold diameter/parison diameter) is typically 0.5 to 3, but up to 7 or greater is possible with some resins.

Some examples of blow-molded products are shown in Figure 4.15.

## 4.4 Injection Blow Molding and Stretch Blow Molding

Injection blow molding is a combination of the extrusion blow molding process and the injection molding process. In this process an injection-molded preform is first made, which is later reheated and then blown into the final finished article. The advantage of this process is that we have the opportunity to form sharp geometric features such as threads

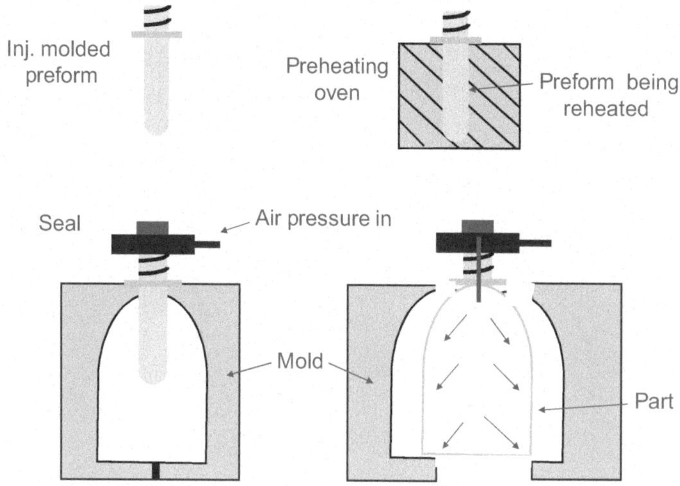


**Figure 4.15A** Injection blow molding process.

and can produce walls of varying thicknesses. This is illustrated in Figure 4.15A [16].

The process starts with injection molding the preform in step 1. In step 2 the preform is reheated in an oven until it is soft enough to be reformed. In step 3, we transfer the warmed preform into a mold and as in extrusion blow molding we use air pressure to expand the heated preform to conform to the mold cavity. The part is cooled in the mold and then the mold is opened and the part is removed. Another preform will be placed in the mold and the cycle repeated.

The advantage for the designer of this process is that the precision of injection molding is combined with the ability of the blow molding process to make hollow articles. The most common use of this increased precision is in the manufacture of bottles with threads that can be reopened many times and effectively sealing bottles from leaks. Examples of this would be PET soft drink bottles and PVC oil containers.

Design considerations are basically similar to those of extrusion blow molding with the exception that geometric features, usually limited to the top of the container, can be produced and wall thicknesses can be varied when you manufacture the preform and carried forward into the final blown article.

In stretch blow molding, as illustrated in Figure 4.15B, a stretch rod or plunger is added to the process that mechanically stretches the preform as it is being blown. The air valve to provide air for blowing is generally located within the stretch rod. This is especially useful when

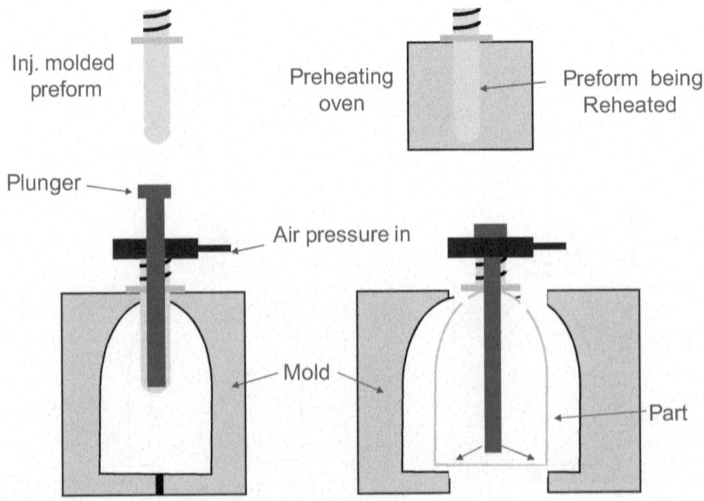


**Figure 4.15B** Injection stretch blow molding process.

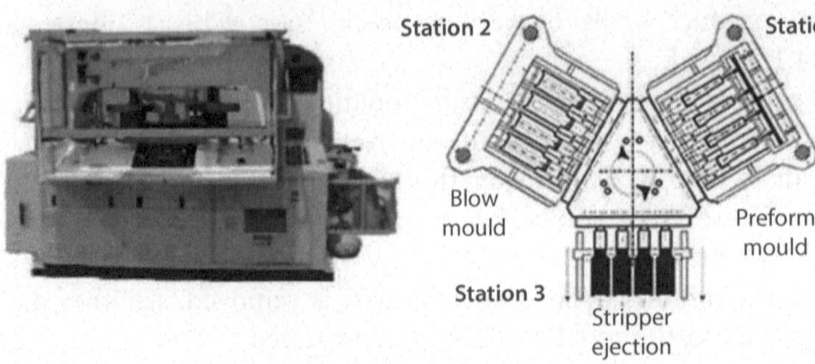


**Figure 4.16** Injection blow molding combined machine center.

trying to produce bottles or containers that are relatively long in relation to their diameter. This is a very common process now being used for water and soft drink bottles.

This process can also be combined into a single operation with a combination injection-blow molding machine, as illustrated in Figure 4.16.

The advantage of the combined process is a generally lower unit cost and elimination of the need to reheat and handle the preforms. The disadvantages are a more complex combined process, a higher capital cost and less process flexibility. It is an excellent choice for very high volume production.

## 4.5 Compression Molding

Compression molding is a cyclical, discrete process that can make simple moderately complex parts from thermosetting polymers. Although primarily used for thermosets, compression molding can be used for thermoplastics for sample preparation or making a low stress molded part.

Figure 4.17 shows a schematic of the process. The molding process consists of placing a charge of material in an open die (step 1), closing the mold halves and forcing the material under heat and pressure to flow and fill out the mold cavity and cure, solidifying the part in the mold (step 2), then opening the mold and ejecting the part (step 3), and repeating the cycle if more parts are required. This process is also called matched die molding. The compression molding process requires relatively low pressures and simple molds, but items produced require trimming of the flash around the periphery of the part where the mold halves come together.

Design considerations include using generous radii and fillets and being able to accommodate a generous parting line where the material flashes from the mold. The parts will have to be drafted generously on both inside and outside surfaces. Another important consideration is

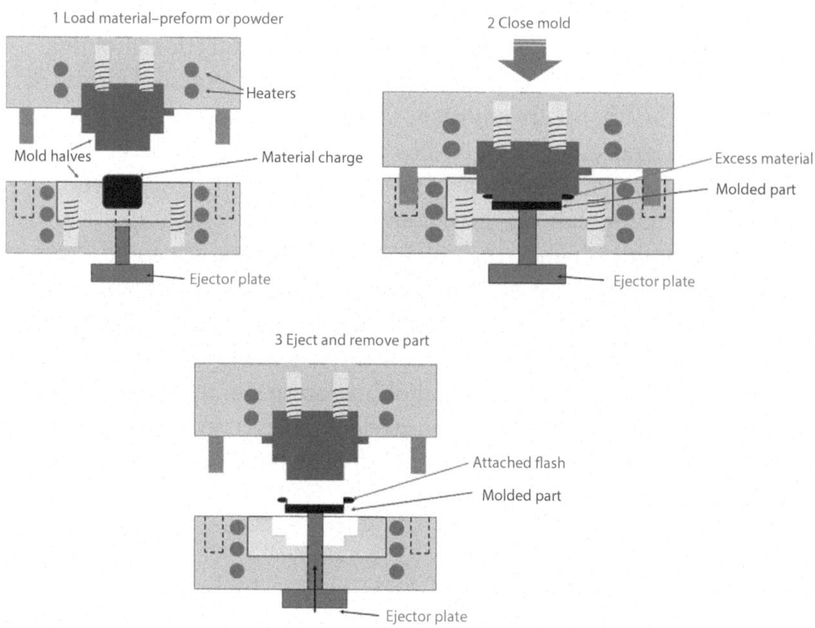


**Figure 4.17** Compression molding process.

that part ejectors need to be balanced with enough surface area to prevent cracking of the part upon ejection. This will leave a number of marks from the ejectors on one surface of the part.

Compression molding will work with almost any polymer that can be made to flow, including both thermosets (including rubbers) and thermoplastics. The process is particularly well suited for highly filled, hard to flow materials, including long glass and carbon fibers and glass fabrics or mats.

## 4.6 Transfer Molding

Transfer molding is a process for molding thermoset materials. Like compression molding, it is a cyclical, discrete process, but it produces net shape parts that do not require any trimming with the exception of gate removal. Figure 4.18 shows a schematic of the process.

In step 1 a material charge, usually in the form of a cylindrical shaped pill, is placed in the transfer chamber (transfer pot). One charge can fill multiple cavities. The material charge is heated in the transfer chamber.

In step 2, after the material has been adequately heated, it is forced by a hydraulic ram through a runner system and gate into the part cavity(ies). This is analogous to the injection cycle in injection molding.

In step 3 further heating cures the part (forms crosslinks) and it is ejected from the mold. In addition to the parts, you also get a cured runner system and cull, which is basically waste and must be disposed of. Some of this might be ground and used as filler in other parts.

Transfer molding can produce complex parts economically. The tool costs will be a little higher than that for a compression mold. Because the material is moved through a runner system and gates, this process is not as good as compression molding for materials reinforced with long fibers, as this movement through the gates will damage the fibers. Design considerations for transfer molding are generally similar to compression molding. Additionally, the designer needs to consider the size and location of the gate and the effect it will have on the part appearance and the stress in the part.

## 4.7 Rotational Molding

Rotational molding is a cyclical, discrete process that is ideally suited for making large to very large seamless, hollow parts, including

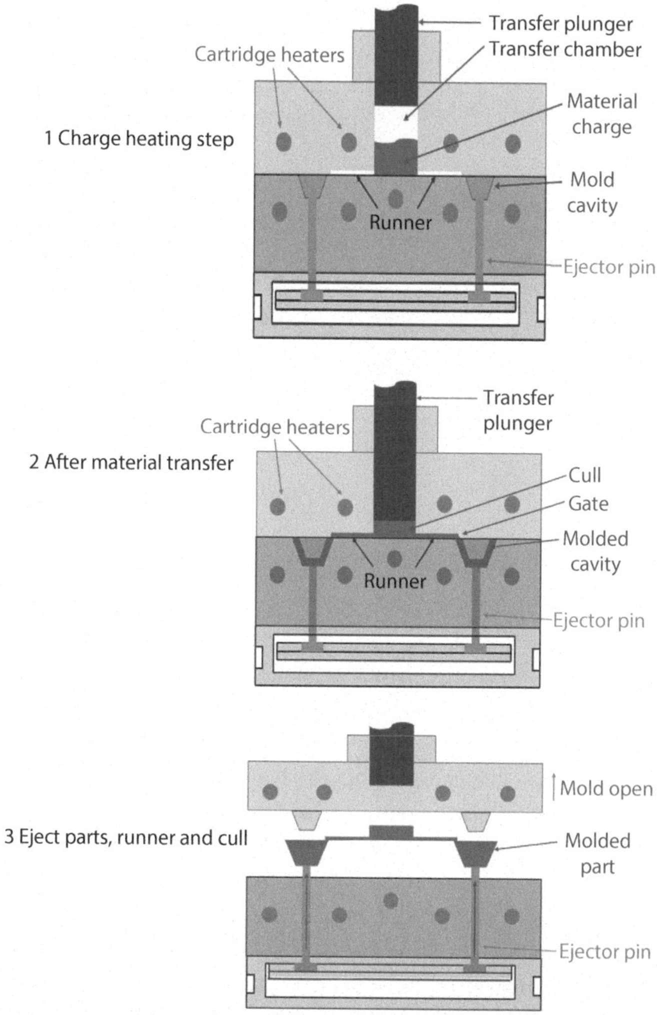


**Figure 4.18** Transfer molding.

cylinders or spheres. The rotational molding process is also called rotomolding or rotational casting. It is a zero pressure process that produces parts that are essentially stress free.

The rotational molding process is depicted in Figure 4.19. In the first step at the load/unload station a precisely weighed charge of material, usually a fine ground thermoplastic powder or a vinyl plastisol, is manually loaded into a mold and the mold is closed. At this point the mold is closed and the machine will index the mold to the oven and start to

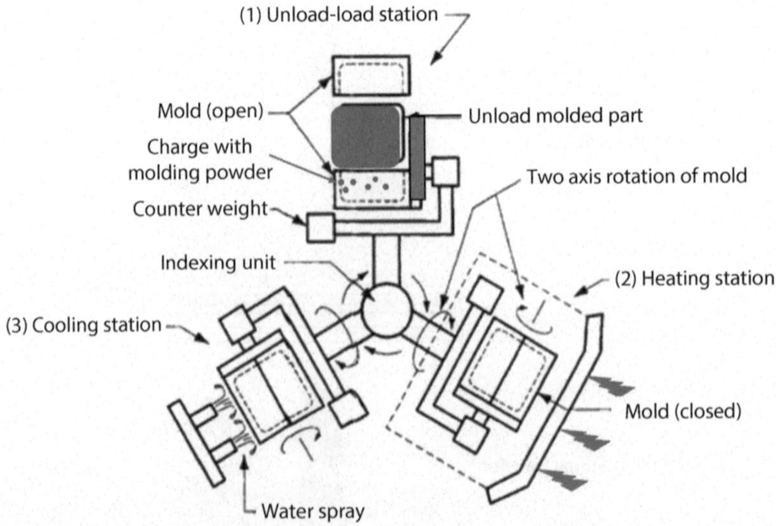


**Figure 4.19** Rotational molding process.

biaxially rotate the mold. As the mold heats, the powder melts and is evenly distributed over the inner surface of the mold by the rotation. The warm plastic becomes sticky and starts to adhere to the mold walls. With continued heat and rotation all of the particles join together. This is called fusion or sintering. This is the longest part of the process. Cycle times can take from several minutes to up to an hour or more for a very thick part. The mold, still rotating, is indexed into the cooling chamber. Here the mold is cooled by compressed air, a water spray or a combination of both. After the mold (and part) have cooled, the mold is indexed back to the load/unload station where the mold is opened up and the part is removed.

The majority of rotomolded parts are made out of polyethylene, but a wide variety of thermoplastics and even some thermosets can be used. Materials have to be thermally stable to endure the long cycle times and must be capable of being ground to a fine powder or put into a liquid solution. Particle size must be uniform and MW and MWD must be tightly controlled.

Design requirements include generous radii and fillets. Large flat sections should be avoided. Tolerances are somewhat limited and wall thickness can be controlled to approximately +/− 10%. External rounded threads can be made, sharp or "V" threads should be avoided.

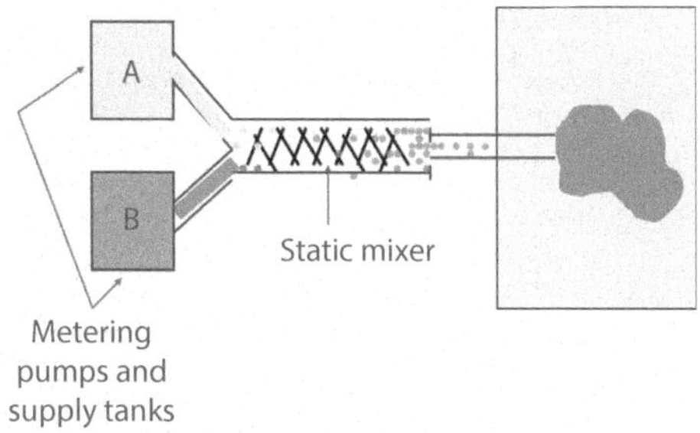


**Figure 4.20** Reaction injection molding (RIM).

## 4.8 Reaction Injection Molding

Reaction injection molding (RIM) is a process that reacts a thermoset material within the mold cavity. RIM normally uses a two-part liquid reactive system, commonly called an "A" part and "B" part resin. Figure 4.20 shows a simplified schematic of the process. Materials are pumped through a mixing head into a mold cavity where they react with each other and crosslink. Reaction times are relatively short and sometimes an inhibitor has to be used to slow the reaction to allow enough time to adequately fill out the cavity.

Because the materials start as liquids, they can flow long distances at very low pressures. This makes the process ideal for making very large parts such as auto bumpers and tractor side panels.

Polyurethanes (both solid and foam) and polydicyclopentadiene are some of the most common materials used in the RIM process.

Shrinkage of these materials can be significant and must be considered in any design. Gates and ejectors are generally small and can usually be easily incorporated into design requirements.

## 4.9 Thermoforming

Thermoforming, also sometimes called vacuum forming, is a process used to shape flat thermoplastic sheets and films into discrete parts. Figure 4.21 shows a schematic of the thermoforming process.

In step 1, a thermoplastic sheet is clamped in a frame and moved under a heating element. The mold underneath the sheet has numerous small holes drilled through it so a vacuum can be pulled on the mold surface. At this point in the process the vacuum valve is turned off.

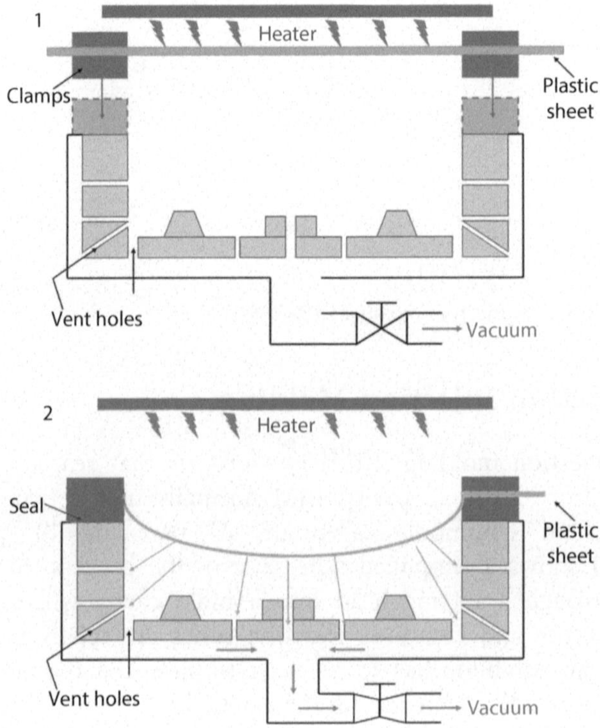


**Figure 4.21** Vacuum forming process.

In step 2, the plastic sheet and clamps are moved on top of the mold to create a sealed box. The plastic has softened due to the proximity to the heating elements and is starting to sag. At this point in the process the vacuum valve is opened up and the vacuum in the mold will draw the softened sheet to the shape of the mold underneath it. The heaters are rolled out of the way, letting the part cool. The part is unclamped, removed from the mold and the process is ready to be repeated.

Sometimes with deep draw or complex parts the sheet will be forced against the contours of a mold by mechanical means. This is called plug-assist thermoforming. The vacuum forming process is unique in that the plastic is softened, but not melted.

Because this is a low pressure process, tooling can be very inexpensive, made of plastic or wood. The low tooling cost makes this process very useful for prototyping.

The process is good to manufacture relatively simple parts with an open or hollow structure. Thicknesses up to ½" are possible, but are typically much thinner. Small undercuts can be made by taking advantage of the viscoelastic behavior of plastics. Again the part design should have generous radii and fillets. While almost any thermoplastic can be used, amorphous thermoplastics work best. For normal vacuum forming, the draw ratio (the depth of the part/width of the part) should not exceed 2:1. For plug-assist vacuum forming the draw ratio can be as high as 5:1.

A modification of this process is the positive-air-pressure thermoforming process. A pressure box is incorporated around the mold area of the machine and 20 to 50 psi compressed air is used to force the molten sheet into the contours of the mold, which gives parts improved detail and structural integrity and closer tolerances along with an injection-molded appearance. Normal vacuum forming uses only atmospheric pressure (14.7 psi) which does not produce as much part detail. Applying pressure on the sheet gives better heat transfer which allows shorter cycle times. In 1975, resin maker Shell Chemical and machine maker Illig of Germany developed a solid-phase pressure forming process (SPPF) to thermoform polypropylene (PP). PP lacks sufficient melt strength for good formability with normal vacuum forming processes. It also is relatively slow to melt and to solidify because of its high crystallinity. The solid-phase PP pressure forming process runs the material colder (below its crystalline melting point) and at a higher pressure (100 psi) [17].

## 4.10 Filament Winding

The filament winding process is a method to produce polymer composites reinforced with continuous fibers. In this process continuous filaments (fibers) are drawn through a bath of liquid polymer and wound onto a mandrel. The mandrel will determine the internal shape of the part. Figure 4.22 is a schematic that illustrates this process. The fibers are drawn into the process from a supply called a creel. The fibers pass through a resin bath and pass through nip rolls, which help smooth out and impregnate the resin into the fiber. The resin-coated fiber is wrapped on a rotating mandrel. As the mandrel rotates, the carriage moves parallel to the mandrel, distributing the fibers uniformly across the mandrel surface. The relative speed of the carriage and the rotation of the mandrel will determine the wrap angle, which will be a determinant of the physical properties of the finished part. By varying the carriage speed and rotation, different wrap angles can be developed on each layer. While helical windings are most common, some machines

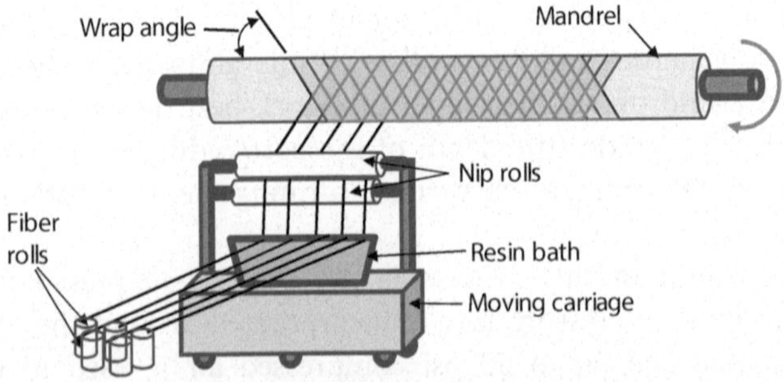


**Figure 4.22** Schematic of filament winding process.

**Figure 4.23** Filament wound pipe.

can apply a circumferential winding with a near zero helix angle or a polar winding with a near 90° helix angle. After winding, the part will be moved to an oven or autoclave to be cured.

Resins used for this process are liquid thermosets. The most common are polyesters and vinyl esters, which are used with glass fibers to produce fiberglass reinforced pipe (FRP), chemical storage tanks and pressure vessels. Epoxy resins are used with carbon fibers to produce a wide range of structural components, including automobile parts, such as the BMW i3 electric car frame, and in aircraft parts such as the fuselage for the Boeing Dreamliner®. Aramid fibers coated with epoxy resins are used in high heat applications.

Care must be taken in handling the fibers during the wrapping process to prevent damage, which will have adverse effects on the physical properties of the finished part.

Figure 4.23 shows a glass fiber/vinyl ester filament wound pipe on top of a plastic core. The picture clearly shows the helical winding pattern.

## 4.11 Pultrusion

Pultrusion is another process that produces polymer composites reinforced with continuous fibers. As in filament winding, continuous filaments (fibers) are drawn through a bath of liquid polymer, but are drawn through a heated die that will determine the internal shape of the part. Figure 4.24 is a schematic that illustrates this process. The fibers are drawn off rolls through a resin bath where they are impregnated with a liquid thermoset resin. The resin-coated fibers are then pulled through a heated forming and curing die where the part is shaped and the resin is cured. Downstream from the die is the mechanical puller which pulls the fibers and parts through the process. After the puller the product is cut to length by the cut-off saw. A surface veil (a fine fiberglass mesh fabric) can be incorporated into the product as shown. Also, in addition to glass fibers, glass mat can be used in the process.

The pultrusion process is a low cost production method that can produce a variety of composite parts. Precise control of actual fiber placement within the composite is difficult. Polyester and fiberglass are the most common materials used but other materials such as epoxies and carbon fiber can be used.

Pultruded parts have applications in the construction and chemical industry where different pultruded parts are used in the form of gratings, ladders, I-beams, U- and L-channels, rebar and customized profiles.

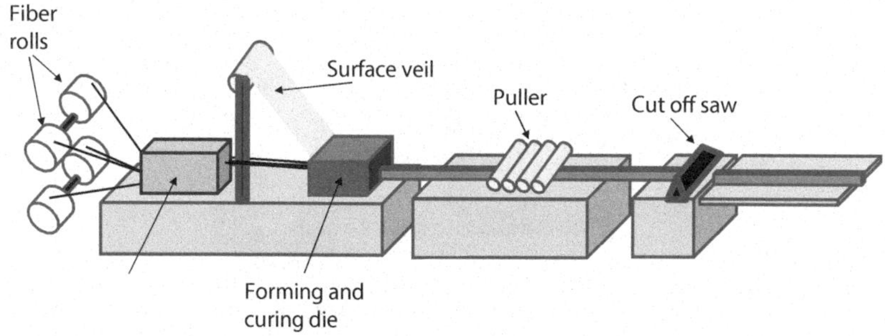


**Figure 4.24** Schematic of pultrusion process.

## 4.12 Additive Manufacturing (3D Printing)

Additive manufacturing was once used almost exclusively for prototyping, but because of its ability to make parts in one piece that cannot be made in any other way and its zero tooling cost, additive manufacturing is finding use more and more as a production process.

Additive manufacturing actually is several different processes that use a 3D CAD file to deposit thin layers of material one at a time to make a solid 3D model of the drawing. The process dates back to 1984 when stereolithography (SLA) was introduced. In stereolithography a laser cures a thin layer of photopolymer that duplicates a small slice of the CAD drawing and repeats the process until the part is complete. The process can produce very fine details but the material selection is limited and parts tend to be brittle. It is a good choice for making sight models.

Fused deposition modeling (FDM) has become synonymous with 3D printing. Figure 4.25 shows a schematic of the process. A thin plastic filament is fed into a fine extrusion nozzle. The X-Y-Z table moves in sync with the cross-sectionally sliced CAD file and thin layer of melted plastic filament is laid down on the platform. The table indexes in the Z axis and the process repeats.

FDM equipment has become quite inexpensive and is easy to use. The process is relatively fast (a few hours or more depending on size), and can come close to design intent mechanical properties. There are many good material choices (ABS, polycarbonate, PLA, nylon, etc.). The finished parts have a relatively rough, sandy surface, but this can be mitigated by using a fine machine setting, which will increase cycle times. Figure 4.26 shows sample parts made by the SLA process on the left and by the FDM process on the right. As can be seen, the SLA part has a smoother surface and reproduced very fine details. The FDM part

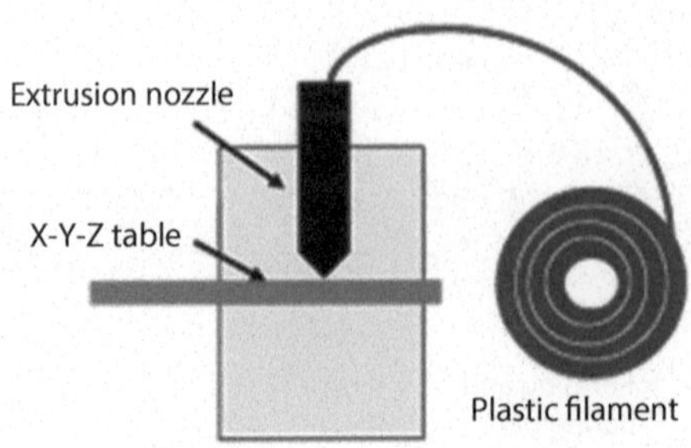


**Figure 4.25** Fused deposition modeling.

**Figure 4.26** SLA part (left) and FDM part (right).

**Figure 4.27** 3D printer (FDM) from Makerbot Industries.

surface is a little rougher, although the part could be sanded if a smoother appearance was required.

Figure 4.27 shows an example of an FDM machine.

The last additive process we will discuss is selective laser sintering (SLS). In SLS a thin layer of thermoplastic powder is positioned on a table and selectively melted and fused. The table indexes in the Z direction and the process repeats. This process is of particular interest because it will also work with metals and ceramics. It is good for making large parts and has a good material selection. Almost any material that can be ground into a fine powder and can be melted and fused will work. The finished part will have a rougher surface than FDM or SLA and might have to be sanded or polished, depending on intent of use, particularly if it will be used for a visual sight model.

## 4.13 Other Prototyping Processes

Additive manufacturing is now widely used because of its speed and ease, but there are several other techniques that are also used extensively.

The most common is machining the prototype from plastic bar or sheet material. Most plastics can be sawed, drilled, milled and turned using conventional tool shop equipment. Most generic materials are available in sheet and bar form. Since the frictional heat generated from these operations can be an issue, slower feeds and speeds are generally used and cutting tools are designed with adequate clearance to rapidly remove the cutting chip. Parts will generally need to be cooled with air, water, a water/soap mixture or a water-soluble oil. Using a CAD program to feed a CNC lathe or mill, gives similar rapid part production and yields a part similar to the design in form and function. The machined parts will be different from production molded parts due to material property variations from the actual production material and lack of orientation and stresses due to molding. Parts will be good to test form and fit, but will only serve as an approximation for function. Machining is a good alternative for very thick parts. Machined parts can also be glued, screwed, welded or cemented together to quickly make a larger assembly.

Casting can be a good alternative when a number of prototypes are required for noncritical uses such as marketing studies, tool design and quotation assistance, or fit tests. In casting, a RTV silicon mold is made by casting around a master made by machining or 3D printing. The mold is split to form a parting line. Next a fill hold is added and the mold put back together. Using a liquid thermoset resin, such as a thermoset polyurethane casting resin, an epoxy, or a thermosetting polyester, multiple parts can be cast from the same mold. For simple parts 50 or more copies can be made. For more complex parts the mold will produce less, possibly as few as 3–5. If a part has some very thin or complex sections, these may be inserted in the mold, as those areas will deteriorate first and the mold life can be extended by replacing the inserts instead of the whole mold.

There are companies that specialize in providing injection-molded rapid prototypes. This is done utilizing a combination of CNC and 3D printing techniques to quickly produce a molded part. This is generally the best way to get a production intent part, but depending on the part it will typically take days to weeks as opposed to hours from other prototyping methods.

# 5

# General Design Considerations

Although each plastic material, process and part will be different, when designing with plastics there are a number of general design considerations that are pretty much universal. Material shrinkage, tolerances, draft and surface appearance have to be considered for pretty much any design. Additionally, sinks and rib design, weld and flow lines, weathering, chemical resistance and environmental factors need to be considered on almost all product designs.

## 5.1 Shrinkage

The first consideration that the designer will face, regardless of material or process, is that of shrinkage. When plastics are heated up and melted, as they cool and harden, as with thermoplastics, or as they cure and harden and then cool down, as with thermosets, the cooling material will contract. As illustrated in Figure 5.1, shrinkage will be three-dimensional. A cooling part will shrink in the x, y and z plane. If the shrinkage rate is the same in all three planes, it is said to be isotropic. If the rates are different, it is said to be anisotropic. Some materials

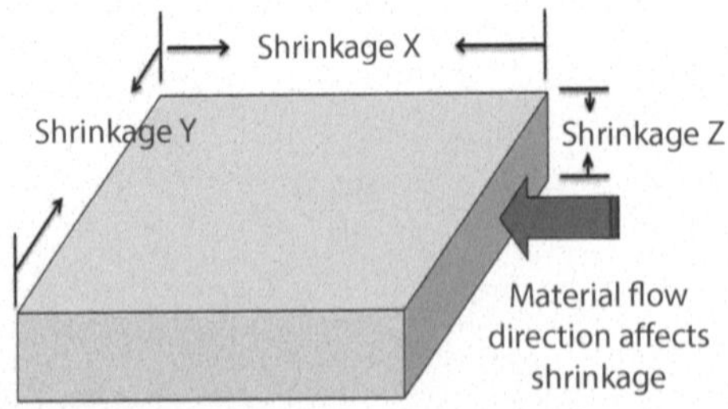


**Figure 5.1** Shrinkage is three-dimensional.

tend to shrink more in the direction the material flows through the tool or die and less across the direction of flow. This tends to be more of a factor in crystalline materials and highly filled materials.

When designing a part, material suppliers will provide data as to what expected shrinkage will be, usually given as a range in inches/inch or mm/mm. On smaller, simpler parts this information will be enough. On larger more complicated parts, if you have previous experience with this material on a similar part, your actual data will be more relevant to you. Without the benefit of this information, when applying a shrinkage range, you want to be sure that if you miss your desired ideal dimension that any correction you might need to make requires taking metal off of your tool or die. This is known as being steel safe. For example, when designing a cup, the material will shrink inward. On the inside diameter, you want to be sure your dimension will be at the nominal or bigger. If adjustment is needed, a simple grinding of the core will fix it. If you were too small, you would have to add metal by plating, welding, or possibly making a new core, all undesirable alternatives. Conversely, on the outside diameter, you want to be sure your dimension will be at the nominal or smaller. If adjustment is needed in this case, again you can make the part bigger by removing metal. Many CAD programs will add in the shrinkage for you, but you have to specify the values to use. Also, there is mold flow software that will predict shrinkage, which can be very useful.

## 5.2 Dimensional Tolerances

Setting the acceptable limits for dimensional tolerances is a critical part of the design process. Tolerances are sometimes confused with shrinkage. Shrinkage is one of the factors involved in tolerances, but many variables contribute to the dimensional stability and achievable tolerances in plastic parts, including the process selected, process variability, material characteristics, part geometry, the part environment after manufacture, and tool design.

Some processes lend themselves to delivering tighter tolerances than others. Higher pressure processes such as injection molding, transfer molding and compression molding can consistently deliver tight tolerances. Lower pressure processes such as vacuum forming and rotational molding will have more variability. Within each of these processes, there will be variability due to the machine used and the quality and repeatability of the process conditions.

Material characteristics will play a significant role in determining the part tolerances that are achievable. Material variability in MW, MWD, and particle or pellet size will affect the finished dimensions. Amorphous materials tend to vary less than crystalline materials.

The part geometry is something the designer has some control over and will influence final tolerances. Parts with wall thickness variations can tend to warp due to differential shrinkage. This can be mitigated somewhat in the process with larger gates and longer cycles, which would increase part cost.

After the part is made, it will continue to shrink for a few hours up to a day or more after it is made. The amount of this post-process shrinkage will depend on the material selected, the process and process conditions. Processes that produce lower levels of stress in the part, like compression molding and rotational molding, will have less dimensional changes after the part is produced. Also, some very hydroscopic materials, such as nylon, will absorb moisture after manufacture, which will change the dimensions. Additionally, the designer has to consider effect of load, temperature, and creep on dimensions when the part is put into service because what is most relevant is the tolerances of the part in functional use.

For the process selected, the tool design can have an effect on the tolerances that can be achieved. Dimensions from features across a mold parting line, such as from a hole produced by a core pin in the fixed or "A" side of the mold to a pin produced in the moving or "B" side of the mold, will have to add the tolerance of the variability of the mold position to other sources of variability. Similarly, dimensions from a feature produced by a mechanical slide or lifter would have some mechanical variability. These can be minimized through good tool design, but must be considered.

Establishing tolerances on a drawing will play a significant role in determining what the part will cost. You need to be sure that the tolerances you specify are adequate so the part will function in any environments you encounter, but over-specification of tolerances can increase product cost significantly. Figure 5.2 shows a simple box lid.

We have dimensioned the part of the lid that fits into the box very tightly. The overall box width essentially fits the air and is appropriately dimensioned quite loosely. Holding one dimension very tightly is usually not that difficult, but as you add more critical dimensions, it becomes significantly more difficult.

How should you go about establishing tolerances? The best approach for setting tolerances is to review similar parts that have been made previously if available. Material suppliers are also a very good resource for their particular materials. I will give some guidelines below, but almost any tolerance can be achieved if you can afford it. On truly super critical tolerances, the last resort will be a secondary machining operation.

The chart (Figure 5.3) is a good general guideline for injection, compression, transfer and blow molding. For vacuum forming a general rule is +/– .015" for the first inch, adding an additional +/– .001" for each subsequent inch [18]. For rotational molding allow +/– 10% on the wall thickness and +/– 5% to 8% on other dimensions, although tighter tolerances can be achieved.

When developing part tolerancing, general good practice includes the following. Try to use low shrinkage materials in parts with tight tolerances. This can be accomplished in many cases with the addition of fillers like glass beads or glass flakes. Avoid tight tolerances in dimensions affected by the alignment of mold halves or moving mold components such as slides. Add design features to lessen the need for tight tolerances such as slotted holes, alignment features and angled lead-ins. Avoid

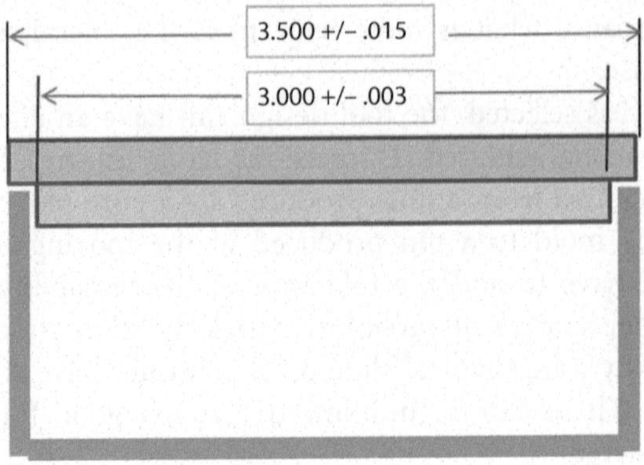


**Figure 5.2** Example of box lid tolerances.

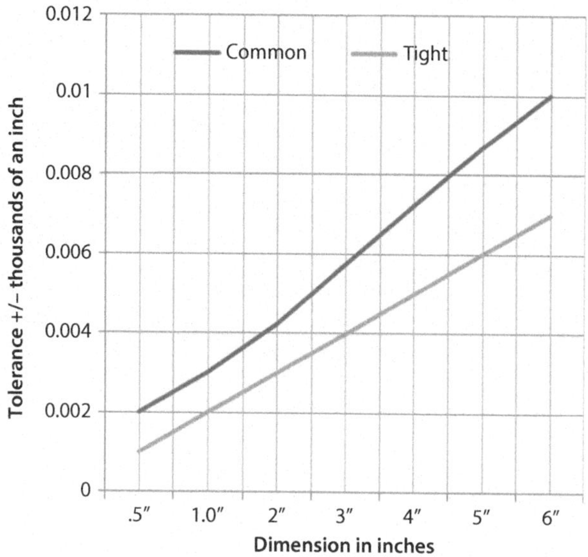


**Figure 5.3** General molding tolerances.

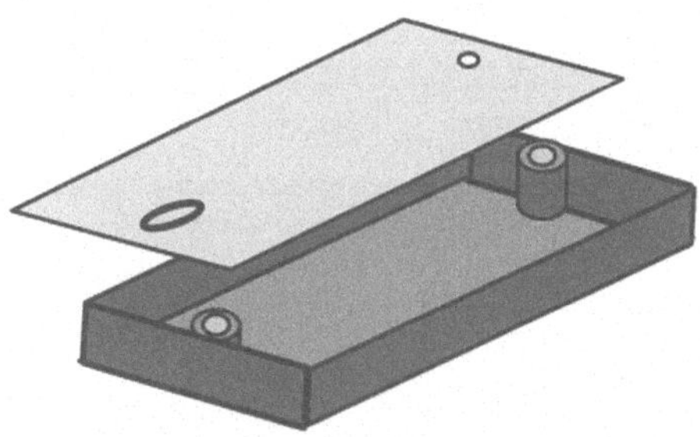

**Figure 5.4** Use of slotted assembly holes.

arbitrarily tight tolerances. In Figure 5.4 for a screw-on box lid, the design has incorporated a slotted hole for one of the lid holes. The one round hole sets the dimension for alignment and can be tightly toleranced. There is no need to tightly tolerance the other hole, and even more importantly, to have a tightly toleranced center hole to center hole dimension.

Figure 5.5 shows some examples of alignment features, including an angled lead-in and a tongue-and-groove alignment.

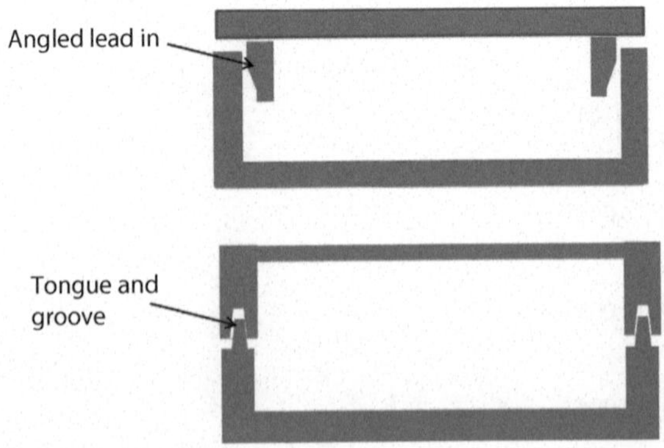


**Figure 5.5** Alignments.

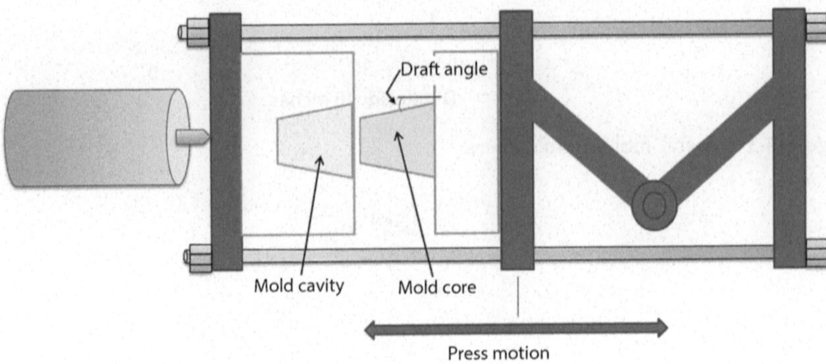


**Figure 5.6** Press motion and part draft.

## 5.3 Draft

When designing a plastic part that will be manufactured by the various molding processes, as the plastic cools it will shrink around the parts of the tool that formed it. In order to remove the plastic part cleanly from the tool without damaging it, the part needs to have some angular relief to facilitate removal. This relief will be required on any design features that are manufactured in the direction of the tool opening or "draw" and any features that are developed by side-pulls, cams or lifters. The concept of draw is illustrated in Figure 5.6.

This is a schematic of an injection press molding a simple cup. The arrow at the bottom of the figure shows the direction the machine, and also the mold, will move to open and close. When the plastic material fills the mold it will shrink tightly around the mold core as it cools. To allow the part to be easily ejected off of the core a draft angle must be put on the core. A zero draft part would be cosmetically damaged or possibly destroyed upon ejection. On the cavity side, you are helped by the shrinkage as the cooling plastic moves away from the cavity wall, but due to the high pressure packing the plastic into the cavity, some draft is also required here, although not as much as for the core side. For draft on a molding core, a minimum of 1/2° per side is generally satisfactory, but 1° is preferable. With very high shrinkage materials more draft might prove necessary. Draft required on cavities can usually be less and often, as little as 1/4° per side can be sufficient, but again, larger tapers of up to 1° help manufacturability. With a very deep draw or very thin part, drafts of 2° to 3° are not unusual. Sometimes draft will be specified as "up to 1° permissible within part tolerance." What this does is reduce the manufacturing tolerance (possibly to zero) and increase the part cost. It is important to recognize that draft is usually a requirement of the manufacturing process and should be incorporated as part of the design. There will be times that the design function absolutely won't allow for any draft. In these cases there are things that can be done in the tool design, such as a collet mold or a mold with 4 side pulls that will accommodate this, but this will be expensive. When these cases arise, the best approach is to get your tool builder involved at the beginning of the design to work with you to concurrently design the part and the tool to develop the best part design/tool design approach.

When you design a feature that requires a side pull or cam, that feature will need to be drafted in the direction of the movement of the side pull or cam much like a core needs to be drafted in the direction of the mold draw. Figure 5.7 shows an example of a typical side pull that is making a hole in a plastic part at 90° to the direction of the mold opening and closing. Again, the plastic will shrink tightly around the pin that makes the side core hole.

When the pin is withdrawn, if it does not have adequate draft, the plastic will stick to the pin and can tear as the pin is withdrawn, or in the worst case, the pin will lock in the part. Draft on side core features should be similar to that used on mold cores.

In compression molding, the draft of the cavity and core permit the material to flow from the bottom of the mold cavity up to the parting line. As the compression mold closes, material flows up the sidewall. This requires maximum cross section at the bottom to insure full material density at the top [19].

For blow molding when producing a round bottle or container, draft is usually not an issue, except possibly at the neck or a handle. Also, if

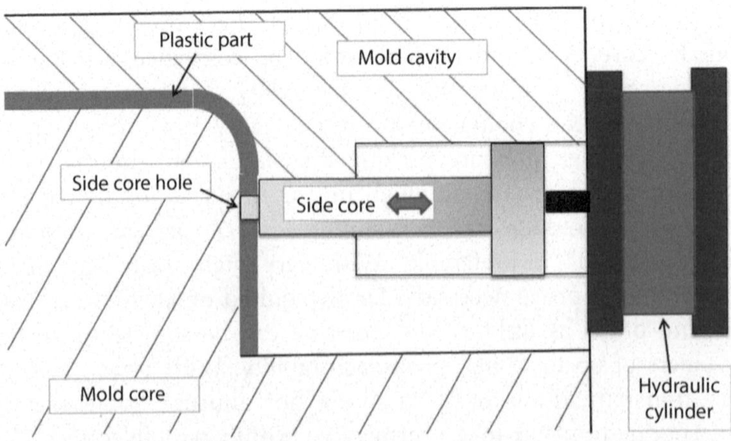


**Figure 5.7** Core pulls and part draft.

the bottle has a texture, a little draft might be required depending on the depth of the texture. For non-round containers, draft of 1–2° will be required.

In rotational molding, a draft of at least 4–6° is required because of lack of an ejection system to remove the part.

## 5.4 Gating

When designing a part for injection molding or transfer molding, the material will enter into the part through an opening called a "gate." This is important to the designer because the gate will leave a mark on the part, so the type and location of the gate or gates will affect the part appearance. Whatever gate type you choose, you will want the gate to enter into the thickest area of the part to insure proper flow and packing of the plastic material as it fills the mold cavity.

The most basic type of gate is the sprue gate. This is when the sprue bushing feeds directly into the part with no runner or gate system. The plastic sprue will come out of the mold attached to the part and must be removed with clippers or a saw. Typically the sprue will be cut from 1/16″ to 1/8″ (1.5–3.2 mm) high (Figure 5.8). The gate can additionally be milled to a closed dimension, but this will add to the cost and the machined surface in sprue area will still be visible.

By using a heated bushing, you can sprue gate without having to remove the cold sprue afterwards. You will still have a vestige from where the plastic enters the part, but it will be glossier than a cut, cold

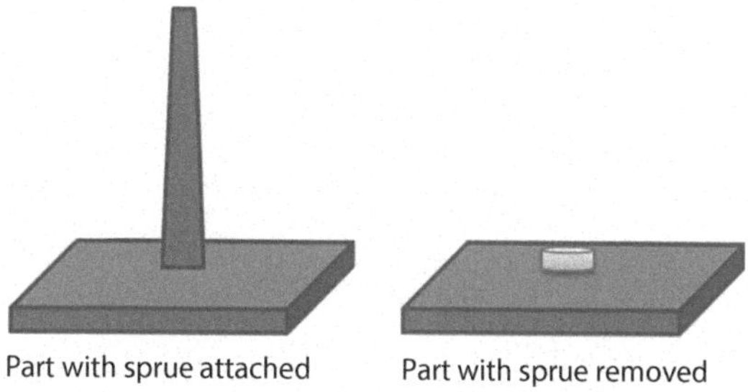


**Figure 5.8** Sprue gate.

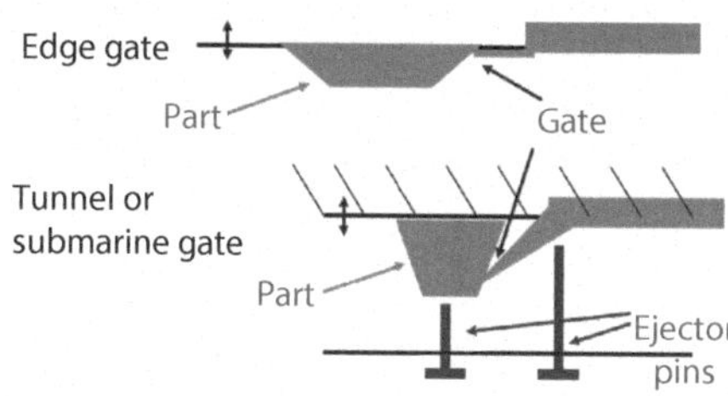


**Figure 5.9** Side gating.

sprue. However, if the design requirement is for the gate to be close to the part, a machining step still might be required.

There are a couple of common ways to gate into the side of the part, shown in Figure 5.9. The edge gate will have to be manually cut and will usually leave either a rectangular, semicircular, or circular vestige on the side of the part. The tunnel or submarine gate cuts itself automatically as the part is ejected from the tool. The sheared gate will leave an oval mark on the part usually less than .010 high or also might tear slightly into the part to about the same dimension as the plastic is sheared.

On round circular or cylindrical parts sometimes an edge gate will cause enough warpage due to differential shrinkage to distort the part more than can be tolerated. A solution to this can be the diaphragm gate shown in Figure 5.10. This gate solves warpage issues but requires a machining operation to remove.

Hot runner gates allow the hot plastic to enter into the cavity directly from a heated manifold. This is similar to a heated sprue but the gate is much smaller. They can be used to gate multi-cavity molds, do not require any secondary operations and work with almost all thermoplastics. They do leave a tiny point of plastic on the part, but this can be hidden by incorporating a small gate well as depicted in Figure 5.11. The size of that point

**Figure 5.10** Diaphragm gate.

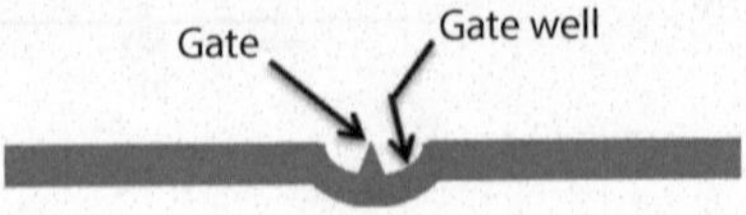


**Figure 5.11** Hot runner gate.

will depend on the diameter of the gate opening and the material chosen, but is usually between 0.010″ to 0.030″.

Hot runners are a little more expensive to build than a conventionally gated mold and a little more difficult to startup, but they are a great solution for high volume production.

The next step up from hot runners are valve gates. Valve gates are essentially hot runners with a piston that seal the gate off. The advantages of this for the designer is that you can have much bigger gates and move more material and the only mark on the part will be a small ring about 0.1 mm (0.003″) high or less, the diameter of the gate. The tool will be slightly more expensive than a hot runner. A valve gate can also be used in place of a cold sprue in a sprue-gated part.

## 5.5 Coring and Holes

When designing a plastic part, the designer should use coring whenever possible to maintain a uniform wall and to avoid excessive thick

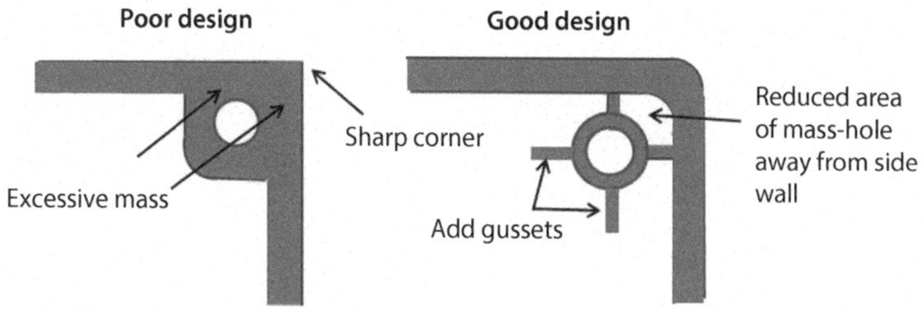


**Figure 5.12** Avoiding excess wall thickness.

sections. This will both help keep costs down and will maximize part strength.

In Figure 5.12 we have two designs for a hole for a shaft in the corner of a housing cover. In the design on the left, the thickness of the areas pointed out as excess mass exceeds that of the nominal wall thickness. This can cause sinks in the exterior wall adjacent to these masses and there is a good possibility that vacuum voids (holes formed in the interior of a plastic part as plastic shrinks in thick sections toward cool outside walls) can form in the thick areas as the part cools, weakening the part. Also, the sharp corner on the outside unnecessarily adds mass. In the design on the right, the hole has been moved out of the side wall and reduced in cross section to 2/3 the nominal wall thickness. The hole itself has been reinforced by adding four gussets to add strength and support to the reduced wall section. Lastly, radii have been added on the inside and outside corner, maintaining the wall thickness through that area at the consistent nominal wall thickness.

In Figure 5.13 we have two designs for a locating post coming off a wall that will attach to a mating part with a snap ring. If you look at the drawing on the left, you will see that the diameter of this post is larger than the nominal wall thickness of the sidewall it is attached to. Also note again the sharp corners where the two walls come together. You should now see how this causes unnecessary thickness in this area. Also, we have a sharp corner where the post meets the wall. This is not a material issue, but can cause stress concentrations which could cause cracking. Since this post is being used to mount to another part, this would be a concern. Now look at the drawing to the right. In the post area a core out has been made in the post to reduce mass. This will be a rectangular cutout going into the post. The basic size and shape of the post has been maintained, but the material thickness has been reduced to no more than nominal wall thickness. Also, we have added

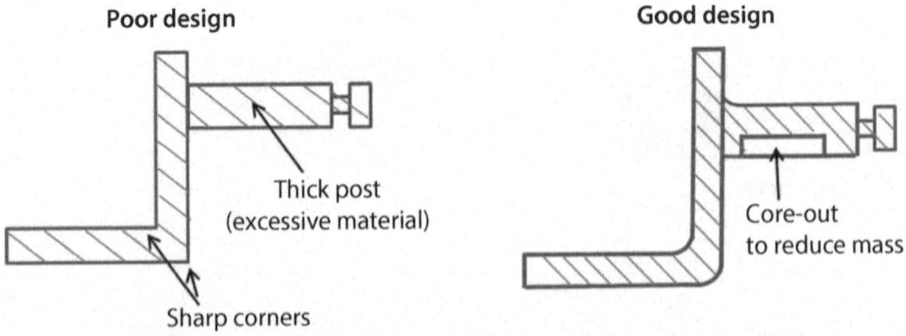


**Figure 5.13** Coring to eliminate excess mass.

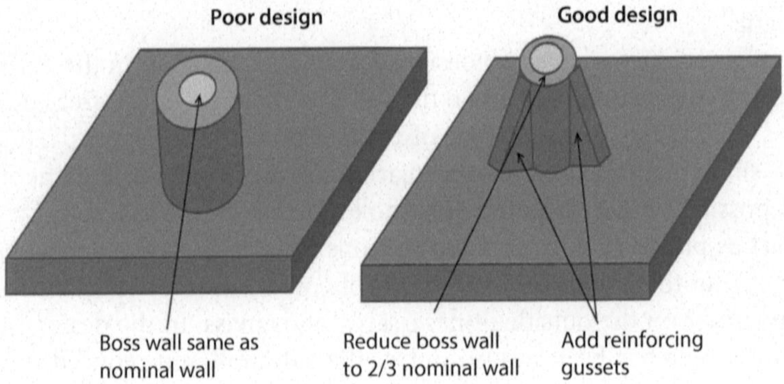


**Figure 5.14** Using gussets to minimize boss thickness.

radii in all the corners to both reduce mass and also to reduce potential areas of stress concentration.

In Figure 5.14 we have two designs for a boss for a self-tapping screw. In the drawing on the left you can see that the wall thickness of the boss is the same as the wall thickness of the base it is sitting on. This will cause a sink behind the boss in the shape of a ring. On the right, the boss has been redesigned to 2/3 of the nominal wall thickness, which will prevent any sinks from occurring on the back of the part. Gussets have again been added to give additional strength to the boss area. When designing bosses for self-tapping screws, you need to look at the screw specifications to insure you have enough material in the boss to retain the screw and balance this with the need to reduce sinks in the part. When you are dealing with a very thin part, it might be impossible to avoid a sink condition and have enough material for a self-tapping screw. A solution to this could be using a molded-in metal inset to engage the screw.

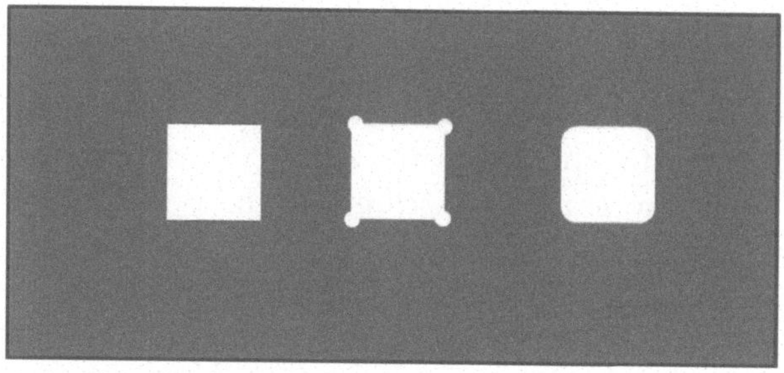

**Figure 5.15** Minimizing stress in square holes.

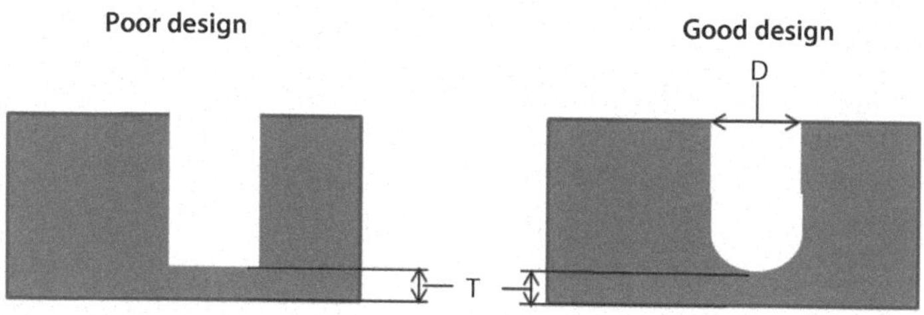


**Figure 5.16** Design for blind holes.

When designing a square hole, sharp corners will act as stress risers and need to be avoided. Figure 5.15 shows a square hole on the left, along with two better design alternatives in the center and on the right.

Figure 5.16 illustrates some designs for a blind hole. In the design on the left, we have a flat-bottomed hole. You should immediately note the potential problem of stress concentration due to the sharp corners at the bottom. At the least, you should put a small radius of at least 0.020″ (0.5 mm). Preferably, the design would be modified as the drawing on the right. Here a full radius is put on the bottom of the hole. Often, with a blind hole, it is not a problem to do this. Lastly, look at the thickness "T" at the bottom of the blind hole. You have to be careful not to get this area too thin or it will be susceptible to break out and will be a difficult area to mold. A good rule of thumb is "T" should be no less than 1/3 the hole diameter [20]. When building the mold, extra care to provide good venting around the pin that makes the hole needs to be taken to ensure that the plastic material will be able to adequately fill this area.

## 5.6 Rib Design

The use of ribs is a common method to improve the rigidity and the strength of a plastic molded part. Used correctly, they will reduce weight, shorten process cycles, eliminate heavy cross sections and improve part appearance. However, ribs should only be used when it is believed that the added structure is necessary to achieve the performance of the part, as they can also cause issues with warpage and with stress concentration if applied improperly.

Figure 5.17 illustrates some basics of good rib design. If you look at the rib on the left-hand side of the drawing, the rib thickness R should be no more than 60% of the wall thickness, N, of the part it will be attached to. If the rib is thicker, the back of the part behind the rib will have a sink mark. When you are adding ribs to a plastic design the rib heights normally are in the direction of draw and therefore will have to have draft added to them. The 60% rule applies to the rib where it intersects the part it is attached to. Geometrically, this will limit how deep the ribs can be. If other design requirements prevent you from adhering to the 60% rule, consider a gas assist process or texturing the back surface, which will mask the appearance of the sink mark.

When your design has ribs that intersect, even if both ribs meet the 60% rule, at the point of intersection they will exceed it and cause a dimple on the surface behind it. On the right side of Figure 5.17 a solution to this problem is illustrated. Stopping the rib short of the intersection will have only a minor impact on part stiffness but will eliminate the sink on the back of the part.

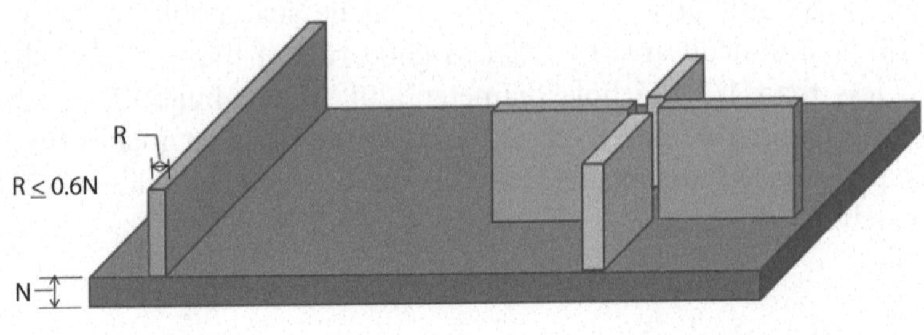


**Figure 5.17** Rib design.

## 5.7 Color and Appearance

Plastic materials can be transparent, translucent or opaque. This is characterized by measuring light transmission and haze. Light transmission is defined as the ratio of incident light that passes through an object to the total transmitted light. Haze is a cloudy appearance that results from forward scattering of light from the sample's surface or from within. Both of these are measured by a spectrophotometer as defined in ASTM test method D1003 [21]. Light transmission is of interest in any optical application and haze is of interest in any appearance application but particularly when designing clear packaging.

Another property of interest for an appearance part is specular gloss. Gloss is defined as the perception by an observer of the shiny appearance of a surface. The measurement of specular gloss consists of comparing the luminous reflectance from a test sample to that from a calibrated standard shiny black surface [22]. The test is defined by ASTM D523.

Plastics can be colored by the addition of a variety of pigments and dyes to give the designer a wide array of color possibilities with a large selection of materials. Characterizing these colors when trying to match different materials or for quality control purposes has always been problematic, but improvements in measurement technologies have made color management a science instead of an art.

The two most common methods to measure color are colorimetry and spectrophotometry. In colorimetry, the quantification of color is based on the three-component theory of color vision. The human eye possesses receptors for three primary colors (red, green and blue), and all colors are seen as mixtures of these three primaries. In colorimetry, these components are referred to as X-Y-Z coordinates. Colorimeters, based on this theory of color perception, use three photocells as receptors to see color in the same way as the human eye sees it. This is the method commonly used for color quality control and some matching. Spectrophotometry is a more precise but more difficult process that uses many more sensors (40 or more in some spectrophotometers) to separate a beam of reflected or transmitted light into its component wavelengths. It measures the spectral reflectance of an object at each wavelength on the visible spectrum. Spectrophotometry is very accurate and repeatable and is generally used for research and color formulation applications.

The appearance of color, haze and gloss are all dependent on the quality of the surface of your plastic part. The surface quality is determined by the material selected, how it is processed, and the surface of the tool, die, or polished rolls used to produce it. One of the important steps in developing a design for a plastic part is determining what the surface finish should be. Tool finishes are commonly specified using the SPI (Society of Plastic Industry) mold finish guidelines. A copy of these guidelines is given in Table 5.1. Previously, a less precise SPE/SPI guideline was widely used and is still referenced occasionally, so a cross reference is included in the last column. The A-1 finish is an optical polish that is used on lenses and mirrors. A-2 and A-3 is a very high polish and is used in applications like automotive exterior parts and high end consumer appliances. B-1 through B-3 are probably the most common finishes used and appear in applications such as computers, printers, most home appliances and hand tools.

**Table 5.1** SPI mold finish guidelines.

| SPI Finish | Guide | Typical Applications | SPE/SPI (old) |
|---|---|---|---|
| A-1 | Grade #3 Diamond | Lens / Mirror – requires 420 SS material | #1 (8000grit) |
| A-2 | Grade #6 Diamond | High Polish parts | |
| A-3 | Grade #15 Diamond | High Polish parts | #2 (1200grit) |
| B-1 | 600 Grid paper | Medium Polish parts | |
| B-2 | 400 Grid paper | Medium Polish | |
| B-3 | 320 Grid paper | Med – Low polish | #3 |
| C-1 | 600 stone | Low Polish parts | |
| C-2 | 400 stone | Low Polish parts | |
| C-3 | 320 stone | Low Polish parts | Approx #4 |
| D-1 | Dry Blast Glass Bead | Satin finish | |
| D-2 | Dry Blast # 240 Oxide | Dull finish | #5 |
| D-3 | Dry Blast # 24 Oxide | Dull finish | #6 |

C-1 through C3 are used in parts where the design intent is a low gloss and D-1 through D-3 are used where a dull or non-reflective surface is desired.

Another option for surface finish is that of texturing. Texturing involves chemical treatment which leaves a distinct pattern on the surface of a mold or a calendering roll which is then transferred to the plastic part. A texture can be very shallow, 0.0001″ (0.0025 mm) deep up to 0.002–0.003″ (0.05–0.08 mm) deep.

When designing for appearance, part of the design plan has to take into consideration the type and location of gates and ejectors as well as any weld lines (also called knit lines) or spider lines that might occur. Weld and spider lines occur wherever two or more melt streams flow together. They will manifest themselves as a colored line on the part to an actual surface separation that will catch a fingernail run over the surface. Weld lines will also be the weakest areas of the part. They can occur from multiple gates, flow around a core pin, a thick section around a thin section or flow through a spider die in extrusion. The best way to manage this is to get the process or tooling engineer involved up front in the design process to look at what the options and tradeoffs are. If multiple gates are required, it might be possible to position them so any weld lines occur in noncritical areas. If gates have to be on an appearance surface for best part properties, consideration can be given to machining or buffing and polishing of the gate area or the use of valve gates to hide the gate as discussed above in the section on gating. Another option can be to use a decorative or instructional label to hide the gate. In extrusion, hollow tubular designs require a spider die, which can cause knit lines. They generally can be kept on the ID surface, but the designer needs to be aware of them.

## 5.8 Chemical Resistance

Plastic materials can have outstanding resistance to chemical attack, making them the material of choice in chemical process piping and laboratory ware, but some of them can be readily attacked (dissolved), which allows them to be solvent cemented. To complicate things for the designer, each plastic responds a little differently to acids, bases, solvents, and oils, so when designing a part it is very important to consider the chemical environment the part might see.

When a plastic material comes in contact with a chemical, basically four things can happen. First, the plastic can be totally resistant to the chemical and will not change at all in its presence. Secondly, the chemical could penetrate the polymer chains, causing the plastic to swell and soften. An example of this would be water softening a polyamide (nylon) material. Third, the chemical can actually separate the polymer chains from one another, causing the plastic to dissolve. An example of this would be an ABS pipe being dissolved by solvent cement. Lastly, the chemical could undergo a chemical reaction with the plastic material. An example of this would be a cellulosic material reacting with an acid. Each of these can also occur to varying degrees.

A plastic material like the nylon mentioned above that absorbs water is said to be hydroscopic. A chemical that is similar in nature to the plastic material will be likely to have a strong interaction. For example, a polar solvent such as water, will be strongly attracted to a polar plastic like nylon. The simple carbon-hydrogen bonds in polyethylene are non-polar and as a result, polyethylene is very resistant to water.

There are a number of tests that are used to evaluate chemical resistance. These can be used to evaluate how a material will respond to a chemical. The first of these is the Immersion Test. The sample is weighed and measured, immersed in a chemical at a defined temperature for a defined time, removed, reweighed and measured, and the overall sample condition is evaluated. This test is described in ASTM D543. In the Water Absorption Test (ASTM D 570) the sample is dried, weighed, immersed in water under prescribed time and temperature conditions and then reweighed. The percent weight gain is recorded. The Stain Resistance Test (ASTM D 2290) is a qualitative test that evaluates a material's susceptibility to staining.

Many material companies and product companies will provide chemical resistance charts that show a material's resistance to a variety of chemicals. When using these it is important to remember that a material that is resistant to two different chemicals separately may not be resistant to the same two chemicals in combination. The only way to be certain would be to perform an immersion test on that combination of chemicals.

In the presence of a mechanical stress and certain solvents, some plastics are susceptible to cracking at the point of the stress. This is known as environmental stress cracking. Environmental stress cracking resistance (ESCR) is measured by ASTM D1693, also known as the "bent strip" test. The test specimens are bent and placed in a holder and immersed in the reagent to be tested. ESCR is the number of hours that 50% of the specimens tested exhibit stress cracks [23].

## 5.9 Weathering and Environmental Effects

Weathering is the degradation of a material from the long-term effects of the environment. Ultraviolet (UV) degradation is the breaking down of a polymer by UV rays (sunlight). The UV light adds thermal energy, which can cause thermal degradation and also can excite covalent bond electrons, making them susceptible to breaking. The physical manifestation of this is a loss in transparency, a yellow or brown color, a chalky appearance and a loss of physical properties. UV absorbers can be added, such as carbon black or benzophenones which preferentially absorb UV, and hindered amine light stabilizers (HALS) which inhibit UV effects by scavenging and quenching.

Some polymers are more susceptible to UV degradation than others. An unmodified polyethylene sheet will degrade to a powder after 2–3 months of exposure to sunlight. An unmodified clear PVC will turn yellow to brown in about 3–6 weeks of direct sun. Polymethyl methacrylate (acrylic) is unaffected after several years' exposure, making it the material of choice for outdoor signs and displays.

Outdoor exposure testing closely reflects natural weathering conditions but requires long exposure times. (There are significant test sites in Florida and Arizona). Accelerated testing using carbon arc or xenon arc lamps can be used to evaluate susceptibility to UV degradation.

Oxidation occurs when an oxygen atom reacts with a polymer. The reasons for plastics to be subject to oxidation are the low energy of bonds in plastic materials, the greater attraction of electrons for a polar oxygen molecule and the relatively open nature of many polymer structures.

Oxygen is readily available and is quite reactive. Oxidation rates increase with increasing temperature. For this reason, plastics are particularly susceptible to oxidation during processing. In extreme cases, processing must be done in the presence of an inert gas to prevent deterioration from oxidation.

Oxidation causes discoloration and a drop in physical properties. Antioxidants, such as hindered phenols and aromatic amines, can be added that very effectively reduce the polymer's susceptibility to oxidation.

## 5.10 Recycling and Recycling Codes

Designers have to be concerned not only with what plastic material or materials will meet the product requirements and give the best value proposition, but also what will happen to them at the end of the product life. This end of life responsibility is a requirement for some markets and companies and is something that needs to be considered throughout the design process. An obvious example of this is plastic bottles, which are required to be recycled by most states or cities. Materials need to be able to be separated and identified when the product is no longer to be used so the materials can be sorted and entered into separate streams to be recycled and reprocessed. With simple parts, such as bottles made of one material, all that is required is that the material the part is made of needs to be permanently identified so it can be sorted and entered into the correct recycle stream after the bottle has been used and discarded.

When parts are more complex, such as assemblies made of parts from different materials, this can be more difficult. In addition to considering how to design the parts to enable easy assembly, commonly referred to as design for assembly (DFA), consideration also needs to be given into how to take them apart. For this purpose, consider self-tapping screws instead of molded-in metal inserts with screws and snap fits that have built in release features to aid disassembly, instead of cementing or welding.

Material identification is usually done using the standard resin identification codes as outlined in ASTM standard D7611, Standard Practice for Coding Plastic. These codes are shown in Figure 5.18. The figure shows the new standard which took effect in 2013 as well as the old convention. The chasing arrows were replaced with the equilateral triangles to eliminate confusion that the arrows indicated the materials

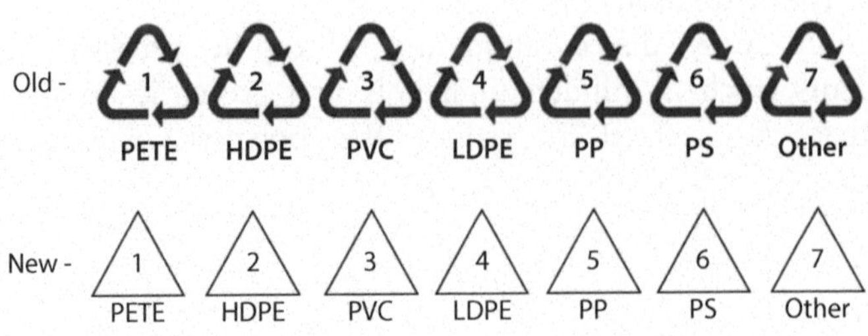


**Figure 5.18** Standard recycle code labels.

were already recycled. Symbols 1–6 cover the highest volume of common recycled plastics and symbol 7 can be used for all other materials. These codes can be incorporated as part of the molding to aid in the recycling process. Materials can be sorted by these codes and products from different manufacturers can be put into the recycle stream.

# PART 2
# PLASTICS PRODUCT DESIGN

Plastic materials provide the designer with the opportunity to create parts that meet functional and appearance requirements that can be difficult or impossible to achieve with other materials, often at a very attractive cost. Although basic engineering principals apply, the product design must take into consideration the unique behavior of plastic materials to yield a part that performs as desired for the life of the product.

This section will go over the engineering basics for designing a plastics part and will also discuss the design of a number of different product types and the special considerations that need to be accounted for with each. We will finish up with a review of joining techniques that are applicable to plastic assemblies and a discussion of the product development and design process.

Some topics that may be applicable to several product groups will be discussed in detail in only one section to prevent repetition. An example of this would be coefficient of friction which is discussed in detail in the chapter on bearings, although it is a consideration in both gears and snap fits.

# 6

# Structural Components

When designing structural components out of plastic, although the basic principles of mechanics apply, because of the viscoelastic behavior of plastics that was discussed in Chapter 2, you need to take several factors into consideration that wouldn't necessarily be a part of a design in metals or ceramics. The absolute magnitude of the loads involved are, of course, important, but you also must consider the duration of the load, the type of load and the environmental conditions that the part under load will be subjected to. Plastics under load will deform over time and will respond differently to instantaneous or continuous loads. Plastics will respond to loads differently as temperatures vary within their operating range. Chemical and weathering factors can also influence how a plastic responds to a load. The less you know about what the range of these influences might be, the larger a safety factor you will want to apply.

In addition to the loads and conditions that the part will see in service, you also need to consider the loads and conditions the part will see in manufacturing, assembly, shipping and storage. Remember the example in Chapter 2 of the pipe designed for a high temperature application that became brittle and broke during shipment in a Minnesota winter! Many times parts are manufactured, stored and shipped in much tougher conditions than they are used in. Assembly operations can impose high stresses on a part (snap fits, self-tapping screws) and decorating operations (paint, coatings) can expose the parts to strong chemicals and solvents.

## 6.1 Rigidity and Strength

Let's first revisit the graph of plastics stress-strain behavior from Chapter 2, shown again in Figure 6.1. Below the proportional limit, an increase in stress applied will result in a proportional increase in strain. Unlike metals, however, the slope of the curve will vary significantly with changes in temperature and, to a lesser extent, changes in loading rate. Due to the viscoelastic nature of plastics the long-term mechanical properties are affected by both time and temperature. As a plastic part is loaded over time the polymer molecules will seek the lowest stress state they can find and they will stretch and move over time to achieve this. Also, as the temperature is increased the molecular vibrations will increase, making it easier for the chains to slide over one another, increasing the amount of deformation that the part will see. This is illustrated in Figure 6.2 where as the temperature increases the slope of the stress-strain curve and, therefore, the modulus decreases. In Figure 6.1 the elastic limit and the yield point are shown as the same point, although the elastic limit can actually occur slightly

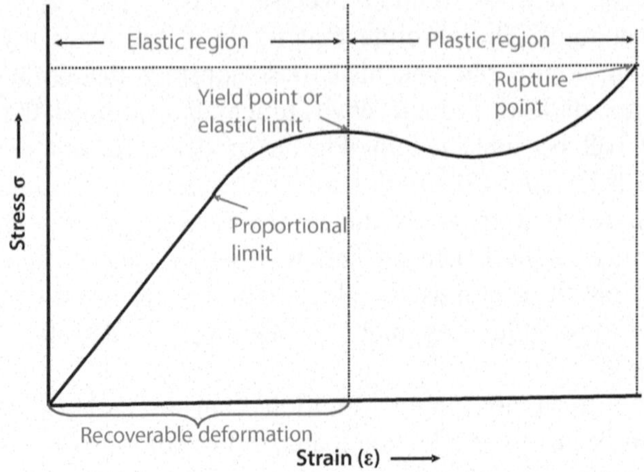


**Figure 6.1** Plastic stress-strain behavior.

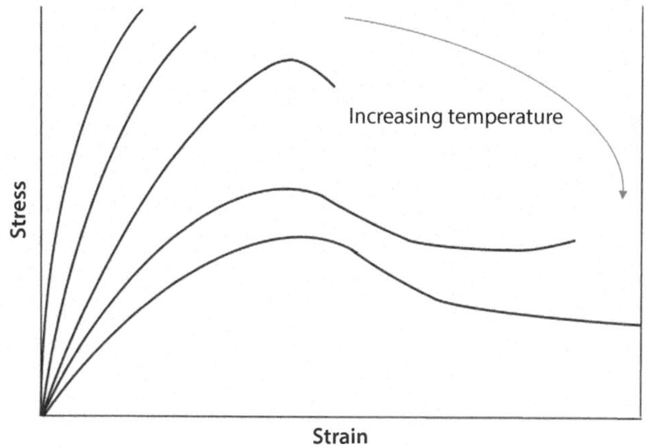


**Figure 6.2** Stress-Strain vs Temperature.

before the yield point. Beyond the elastic limit, the part will incur some permanent deformation that will not be recovered even if the load is completely removed. This deformation is called *permanent set*. At the yield point, the part will continue to deform without any additional stress. At and beyond this point, all deformation is nonrecoverable. If the part function cannot tolerate a permanent deformation, you must stay in the range below the elastic limit.

Again, in Chapter 2, the tensile modulus or Young's modulus was defined as the dividend of dividing the stress by the corresponding strain on the curve below the proportional limit. Often, even at low levels of stress, the curve will not have a clear linear region for the calculation, unlike that shown in Figure 6.1. A solution to this is to calculate a secant modulus from the curve on the stress/strain diagram at the specific point (stress level) you are interested in.

With plastic materials a stress/strain curve will give a modulus at a specific temperature and a specific loading rate. Although limited, the data is useful to obtain an estimate of plastic material's strength and rigidity relative to another material. In the design of parts that will see only momentary or intermittent stresses, such as a case in which a part snap fits together or a handle that will be lifted occasionally, the data would be adequate. These properties do not give information about how the plastic will react to changes in temperature or when subject to a constant load over a period of time. For this information, it is necessary to use creep curves generated from creep data.

## 6.2 Creep

In Chapter 2 we defined creep as an increasing strain over time in the presence of constant stress. When a plastic is subjected to a static load (at a constant temperature) it deforms quickly to a strain that is predicted by a normal stress/strain curve. It then will continue to deform at a slower rate indefinitely. If the load is high enough at some point rupture will occur. Plots of this response are called creep curves and are characterized by an initial stage of rapid deformation, a second much longer stage where deformation continues at a relatively slow rate, and if time is long enough, a third stage in which creep rupture occurs. As predicted in Figure 6.2, if the temperature is increased, the rate of deformation will also increase. Importantly, a plastic material subjected to a *given load and temperature* will deform over time in a very predictable way.

Figure 6.3 shows an idealized series of creep curves at a constant temperature. Strain is plotted against time for a plastic test sample for a series of different constant loads at a constant temperature. The sample subjected to the highest load is represented by the curve farthest to the left. Each succeeding sample is at a lower load. In the first curve the sample is loaded at a stress close to the yield stress that was predicted by a stress/strain curve. The sample quickly goes through the initial and secondary deformation stages and the sample ruptures very quickly, in seconds or within a few minutes. The second sample has a reduced load, and although it goes through deformation and rupture, it takes a

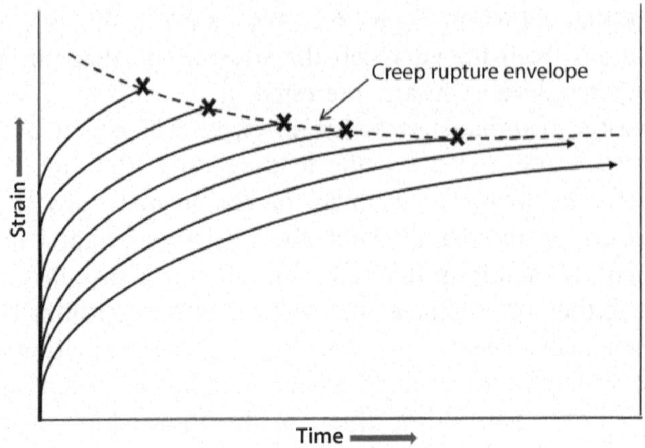


**Figure 6.3** Creep rupture curves.

longer time, many minutes to an hour. As the load is continually reduced on the subsequent samples the time to rupture gets progressively longer until a point is reached where no rupture occurs in the time horizon of the test. This time can be days or years. By connecting the rupture points we can define a creep rupture envelope. This envelope defines the maximum stresses and times that we can safely use the creep curves to predict long-term deformations.

With a continuous applied load the strength of a plastic part can be quite a bit less and the deformation quite a bit more over time than would be predicted by an instantaneous stress-strain curve. Therefore, when designing a part that will see more than an instantaneous or occasional load, creep data must be used to provide the basic design data.

As the creep curves are generated, the majority of the deformation occurs early on, usually in the first 1,000 hours. As the time increases, the curves tend to flatten out. This raw data lends itself to extrapolation and when graphed out on log-log or semi-log paper is normally a straight line. This allows prediction of deformation at times well beyond the limits of the test data. Figure 6.4 shows a typical plot of creep data for an unfilled polycarbonate material at 73 °F. The data points go out to 10,000 hours, about 1.14 years (one year = 8760 hours) but the plot

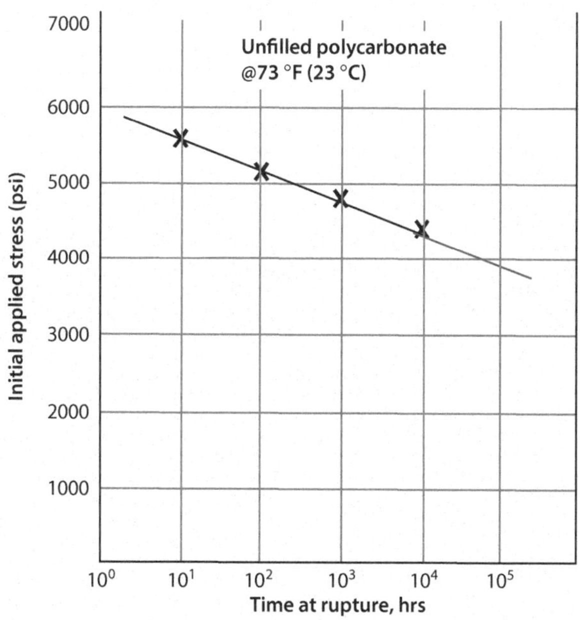


**Figure 6.4** Plot of Stress vs Time at Rupture (creep curve).

on semi-log paper fits the data to a good extrapolable straight line that can be used to predict rupture points at future times.

This data can be obtained from a number of sources. A good place to start might be one of the online databases available, such as IDES Prospector®, which allow access to a wide variety of materials and manufacturers. Alternately, there is published data from a number of sources, such as *Modern Plastics Encyclopedia*, which pioneered the publication of this data back in the 1960s. Finally, as you narrow your material choices down, the individual manufacturers will have data on their specific materials. This is important because there can be differences in the data of similar materials from different manufacturers.

Another very useful form in which creep data is presented is the isochronous stress-strain curve, which plots tensile stress versus strain at a series of different time increments. An example of this plot is given in Figure 6.5. To determine the creep modulus (also called the apparent modulus), divide the calculated stress by the resulting strain that is read from the curve corresponding to the time duration desired. For example, if we have a load of 4000 psi for a period of 1000 ($10^3$) hours we will have a corresponding strain of 2.0%. Dividing the stress by the strain (4000 psi/2.0%) we get an apparent modulus of 200,000 psi. If we look out to 10,000 ($10^4$) hours we find that the strain has increased to

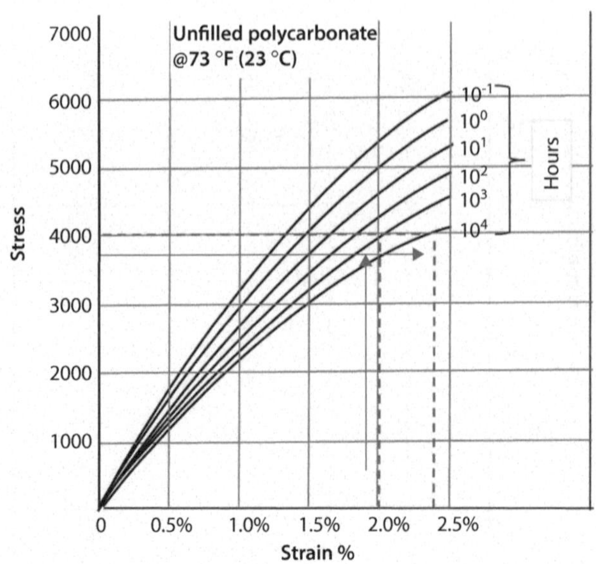


**Figure 6.5** Isochronous stress-strain curves.

2.4% and we calculate the apparent modulus at this point to be 166,667 psi (4000 psi/2.4%). Because of the viscoelastic behavior of plastics, you must use a modulus value appropriate for the time the load will be applied. The apparent modulus enables prediction of deformation over time.

When a plastic structural component is subjected to a force, it will bend in response to that force as any material would. This bending will cause tensile stresses on the convex side of the part and compressive stresses on the concave side of the part. The point at which the stresses change from tension to compression (the point where the stresses are zero) is called the neutral axis. The stress distribution through the part is defined by the flexure formula for beams,

$$S = \frac{Mc}{I} \tag{6.1}$$

where $S$ = the maximum stress, $M$ = the bending moment applied, $c$ = the distance of the neutral axis of the beam to the point of interest, and $I$ = the moment of inertia of the cross section of the structure. The moment of inertia indicates the resistance of a structural component to bending and is dependent on the geometry of the structural component. It is expressed in units of inches$^4$ (SI) or mm$^4$ (metric). To simplify calculations we calculate a section modulus, $Z$, which = $I/c$. This simplifies Equation 6.1 to:

$$S = \frac{M}{Z} \tag{6.2}$$

Figure 6.6 illustrates this concept for a simple rectangular beam and shows calculations for area A, neutral axis c, moment of inertia I, and section modulus Z.

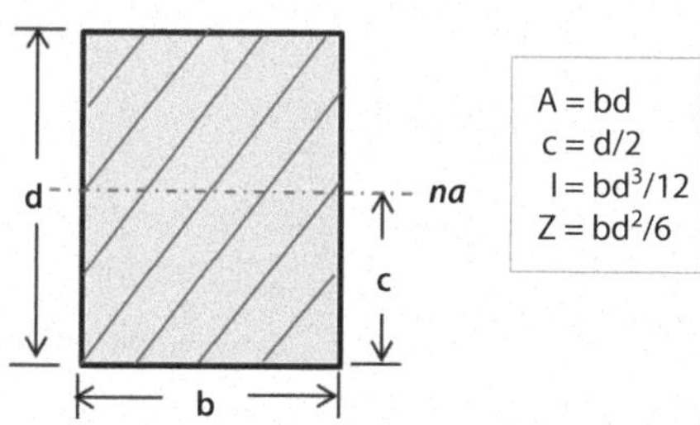


**Figure 6.6** Beam loading problem.

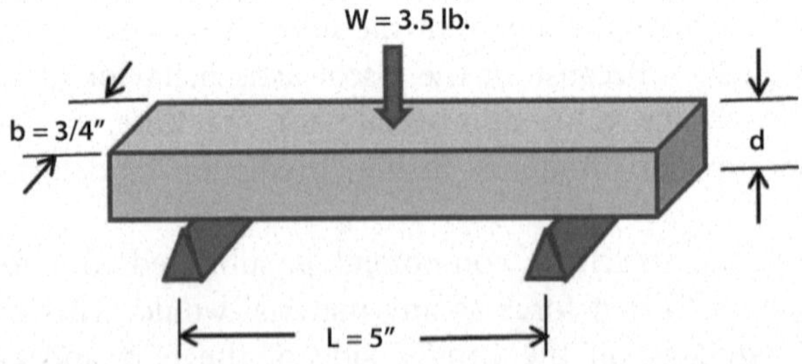


**Figure 6.7** Beam loading problem.

These data for a number of other common geometries are given in Appendix B and many more can be found in a number of mechanical engineering handbooks.

Next, we will look at some examples of using creep data to design components under a structural load for a period of time. When designing a plastic part you need to know not only the load conditions but also at what temperature the load will be applied, and how long the part is expected to function. You always don't have that data set out specifically, so you need to make reasonable, conservative assumptions when deciding on what creep curve to use to develop your data.

Assume we have been asked to design a beam which will support a component in an assembly that will work in an air conditioned environment. We believe the life of the assembly will be no more than 10 years. Figure 6.7 gives a graphic depiction of the beam and the design problem. We are given the length and width of the space the beam will fit into and have to determine how thick the beam must be to meet the design conditions and to support the load required. Also, because of the space restrictions within the assembly, over the course of its 10 year life, no more than 0.100″ of deflection can be tolerated.

To solve this problem we will use creep data curves and the following design equations:

$$S = \frac{3WL}{2bd^2}$$

$$S = \frac{WL^3}{48\Delta \frac{bd^3}{12}} \qquad \text{(6.3) and (6.4)}$$

$S$ = Maximum fiber stress psi

$W$ = Load applied
$L$ = Length of the beam
$b$ = Width of the beam
$d$ = Depth or thickness of the beam
$\Delta$ = Deflection of the beam at midspan

Because we are using polycarbonate in other parts of the design and already have the data, we will take a look to see if this material will work. Using the creep curve from Figure 6.4, which is at the room temperature (73 °F) that the design calls for, we have to extrapolate out to arrive at a design stress for this material at ten years' time, which is equivalent to 87,600 hours. This has been done in Figure 6.8. On the creep curve, we find the point on the X axis of 87,600 hours.

Draw a line vertically to intersect the curve and read off a 10 year design stress of 3750 psi. At this point, it is prudent to consider adding a safety factor and to develop a working stress from the design stress. This safety factor will allow for any variations in service of conditions specified, i.e., time and load, any errors in the data itself as well as an assessment of the risk if the part fails. For purposes of this example, we will use a safety factor of 50%. (We chose this number somewhat arbitrarily, just for demonstration purposes. Safety factors are usually

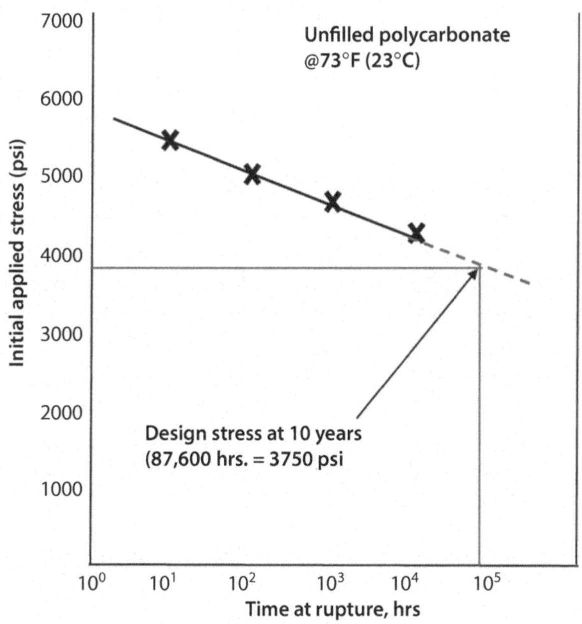


**Figure 6.8** Plot of Stress vs Time at Rupture (creep curve).

determined by codes and standards, accepted industry practice, company standards or a design risk assessment).

Therefore, working stress = design stress × 50% = 3750 psi × 50% = 1875 psi.

From Equation 6.3:

$$S = \frac{3WL}{2bd^2} \text{ and } d = \sqrt{\frac{3WL}{2bS}} \tag{6.5}$$

$$d = \sqrt{\frac{(3)(3.5)(5.0)}{(2)(3/4)(1875)}} = 0.137'' \tag{6.6}$$

The thickness of the beam to meet the stated requirements needs to be 0.137″ to support the load for 10 years without breaking. We still need to look at how much deflection we will have over ten years to see if the 0.137″ thick beam will meet the requirement of no more than 0.1″. We do this using Equation 6.4, the deflection equation.

$$E = \frac{WL^3}{48\Delta\frac{bd^3}{12}} \text{ and } \Delta = \frac{WL^3}{48E\frac{bd^3}{12}} \tag{6.7}$$

From Figure 6.5, the Isochronous Stress Strain curves, we find that at 10 years, for a stress of 3750 psi there is a corresponding strain of 1.8%. Dividing the stress by the strain we find the apparent modulus for this point, 3750 psi/1.2% is 208,333 psi. As with the design stress, we will apply a safety factor to this number. This time we will use 75% and arrive at a working modulus of 156,250 psi. Substituting this in the equation above we find the deflection, $\Delta$.

$$\Delta = \frac{WL^3}{48E\frac{bd^3}{12}}\frac{(3.5)(5)^3}{(48)(208,833)(0.75)(0.15)^3} = 0.363'' \tag{6.8}$$

We find that although the beam will support the load without breaking, over the ten year timespan the deflection will exceed our maximum allowable by over a factor of 3. At this point there are several possibilities to solve this. The first would be to look at a material with a higher modulus, possibly a glass-filled material. Looking at the formula, you would need to increase the modulus by over a factor of 3 to get the desired result. The deflection varies with the cube of the thickness and simply increasing the thickness might be a better solution. By substituting in the deflection equation, we find that increasing the beam thickness to 0.211″ will solve the deflection issue. You would also want to revalidate all of your design assumptions including requirements and safety factors.

Now, let's look at how we might use creep curves to make a material selection. You are asked to design a shelf support beam to hold up a display shelf. The shelf will be 3″ wide and it will see a maximum distributed load of 3 pounds/inch, including the weight of the shelf. The support beam will be fixed on one end and unsupported on the other. To maintain a nice appearance, no more than 1/16″ deflection of the shelf is desired. The display shelf in service will be indoors in an air conditioned store. The display will be short term in nature, no longer than six weeks. You have worked on designs for other parts of this display and have used ABS for other components because of good strength and processability, good gloss and low cost. For simplicity and cost reasons, the molder asked if you can stay with ABS on this part.

Figure 6.9 shows a graphic schematic of the problem. Looking at the problem it is obvious that this is a cantilever beam problem, with the beam fixed at one end, unlike the previous example with support at both ends. The first step in the design process will be to determine the modulus that will be required to meet the strength and deflection requirements that have been specified. For a cantilever beam with a uniformly distributed load, total deflection $\Delta$ is defined in Equation 6.9 [24]. Rearranging to solve for E and utilizing Figure 6.6 for the moment of inertia of a rectangular beam we get Equation 6.10 which we will use to

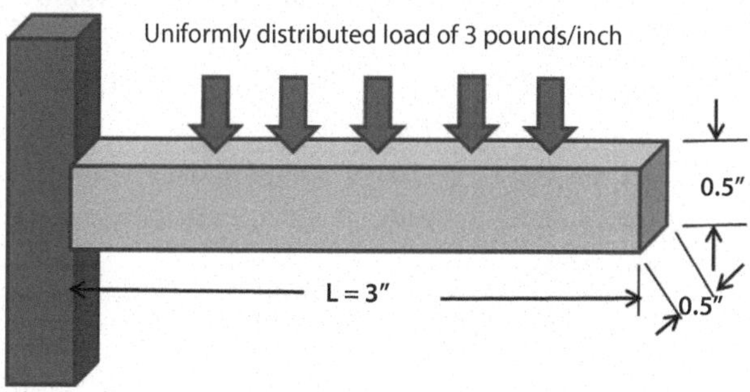


**Figure 6.9** Cantilever beam.

find out what modulus will be required for the design specifications that are given.

$$\Delta = \frac{wL^4}{8EI}$$
$$E = \frac{wL^4}{8\Delta \frac{bd^3}{12}} \qquad (6.9 \text{ and } 6.10)$$

$\Delta$ = Maximum deflection of the beam at end
$w$ = Distributed load applied
$L$ = Length of the beam
$b$ = Width of the beam
$d$ = Depth or thickness of the beam
$E$ = Apparent modulus psi
$I$ = Moment of inertia of beam

$$E = \frac{wL^4}{8\Delta bd^3} \frac{(3.0)(3)^4}{(8)(.0625)(0.5 \times 0.5^4)/12} = 93,312 \text{ psi} \qquad (6.11)$$

We need a design modulus of 93,312 psi to meet design requirements. To arrive at a working modulus we will apply a safety factor of 50% and arrive at a working modulus 186,000 psi.

Next, we need to look at some available creep data for ABS materials [25]. Figure 6.10 shows a plot of the apparent modulus of four different ABS materials for data from one to 1000 hours. When you are using these plots, good practice is not to extrapolate more than one decade beyond the data. This data is through $10^3$ (1000) hours so we can reasonably extrapolate out one more decade to $10^4$ (10,000) hours. Our design specification asked for a six week life, which is equivalent to 1008 hours, so we are well within the limits of our data.

Looking at the data, we find that at 1000 hours the modulus for general purpose ABS (GP ABS) is about 180,000 psi, which is essentially the same as our calculated 186,000 psi requirement, so we could use this material for our beam.

After doing some more market research, the customer believes that the items might be on display for a much longer time, possibly up to a year. A year would be 8760 hours, still within an acceptable extrapolation range for our data. Looking at the chart you will see that the

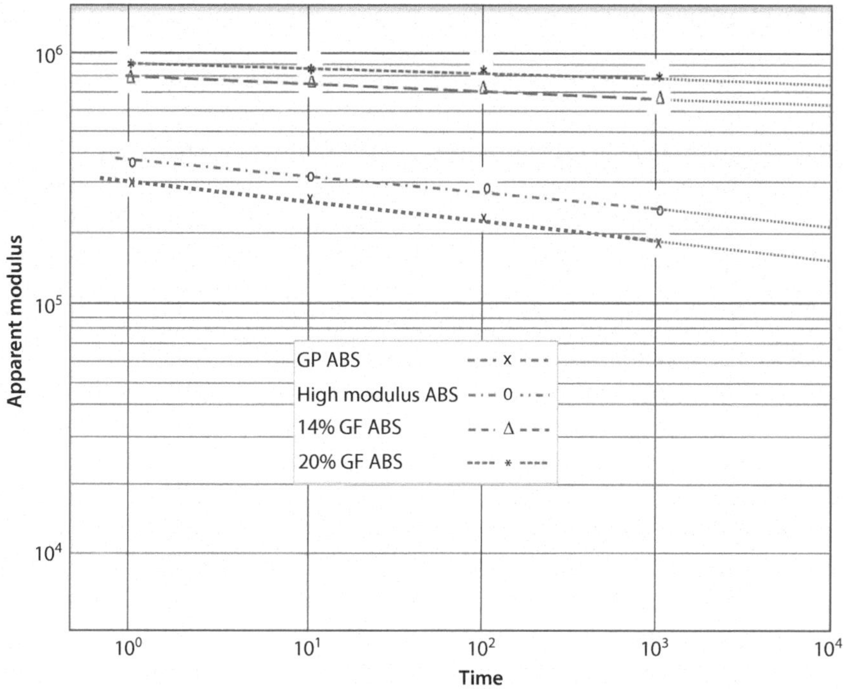


**Figure 6.10** Apparent modules of ABS materials.

modulus for the GP ABS is a little less than 150,000 psi and is too low for the design requirement. Changing the material selection to a slightly more expensive high modulus ABS will give a modulus after a year of a little over 200,000, making it a good choice for the design. When you reach this point, it would be appropriate to explore the creep data for similar materials from different manufacturers to find the best cost/performance solution because using a different grade of a material, from a manufacturing perspective, is the same as using a completely different material.

The important point to take away from this section is that you use the same design equations for plastics that you use for any other material but you must use apparent modulus data derived from creep testing whenever loading is more than instantaneous.

## 6.3 Fatigue

When rigid plastic parts are subjected to cyclical loading they can fail at substantially lower stress and strain levels than the same part would under a static load. Discontinuities in the plastic structure cause localized stress concentrations that under load can cause localized failures within the polymer matrix. These localized failures exhibit themselves as microscopic cracks. Under a cyclic load over a large number of cycles these cracks can grow, ultimately leading to part failure. Products that encounter this type of loading include snap fit closures, integral plastic hinges or a gear tooth.

For both design and material selection for these types of applications, test data that shows the relationship between cyclic loading and life can be very useful. In this kind of testing, a specimen of material at a given temperature is repeatedly subjected to an initially high constant stress at a constant frequency, and the number of cycles to failure is recorded. The stress level is reduced and the test repeated until failure again occurs. This process is continued until the stress is reduced to a level that will generate a life of $10^6$ to $10^7$ cycles. The test data are usually presented as a plot of stress versus log cycles. This stress vs. number of cycles plot is commonly referred to as an S-N curve. Figure 6.11 shows typical S-N curves for some selected plastics. A common test method for developing S-N curves is outlined in ASTM D 671.

When looking at this data it is important to know the temperature that the test was run at as well as the frequency. When plastics are flexed, they heat up in the area of flexure. As the temperature rises, the strength of the material will decline.

The advantage of the S-N curves is that they give an estimate of the number of cycles of a given repetitive stress a part can endure. Some materials will tend to drop off fairly steadily, like polypropylene shown in Figure 6.11, while others, like acrylic, after dropping for a while tend to level off. This leveling off is called an endurance limit. Materials that exhibit this will perform for an indefinitely long period of time when subjected to stresses below this level. In applications where the part will be cycled to a fixed displacement, the load required to produce that displacement can be calculated and the life cycle at that stress level can be determined from the S-N curves. S-N curves are provided by some material suppliers, but sometimes, test data will have to be generated. Extrapolating the results of the S-N data to an actual part has many variables, including geometry, frequency and temperature. The curves are very useful for comparing one material to another, but when

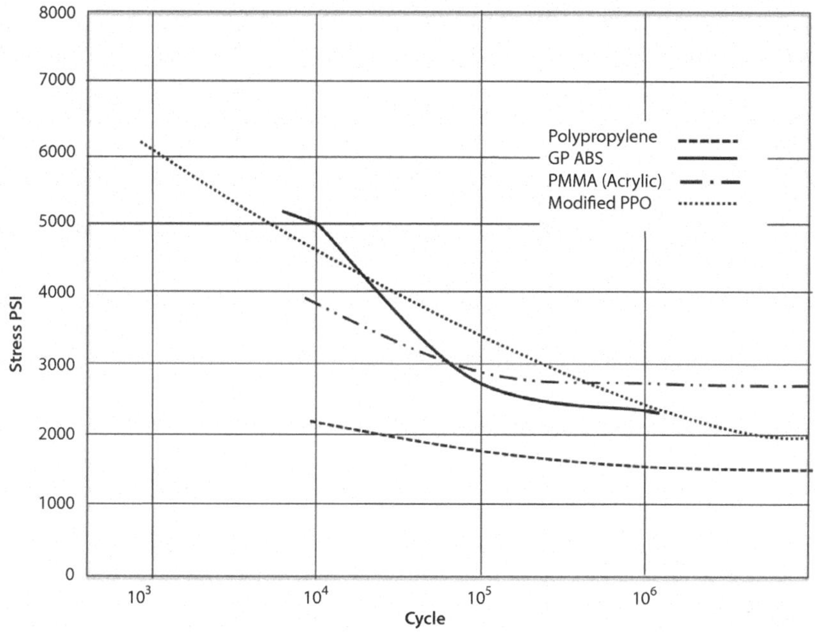


**Figure 6.11** Fatigue Failure Stress vs Cycles (S-N curve).

applying the data to a design, even if the test conditions are very close to the design requirements, a generous safety factor of as much as 3 to 4 times the design stress should be applied. Of course, whenever possible testing of an actual part under design conditions is recommended.

When you are designing a part that will be subjected to a cyclic load, you want to avoid sharp corners, notches, changes in thickness and holes. You also need to try to keep gates and weld lines out of the stressed areas. Plastics reinforced with glass and carbon fibers tend to have better performance in cyclic loading applications.

## 6.4 Torsion

A torsional load occurs when a structural component, such as a round shaft, is subjected to a force that causes it to twist. The equation for the stress in torsion that the shaft will experience is analogous to Equation 6.1 for bending stress S = Mc/I. Torsional stress is identified by the Greek letter $\tau$, the twisting moment is called $T$, $c$ now represents the distance from the center of the shaft to its outer surface and the moment of inertia is replaced by the polar moment of inertia $J$ to yield Equation 6.12, the torsional formula:

$$\tau_{\max} = \frac{T_c}{J} \tag{6.12}$$

The polar moment of inertia, $J$, for a round shaft is given in Equation 6.13:

$$J = \frac{\pi d^4}{32} \tag{6.13}$$

Figure 6.12 gives a schematic of a solid round shaft restrained at one end.

As the shaft is subjected to a torsional stress, it will twist around the center of the shaft. The angle of twist, $\theta$, is the angular amount given in radians that the shaft is deformed. This angle can be calculated from Equation 6.14,

$$\theta = \frac{TL}{JG} \tag{6.14}$$

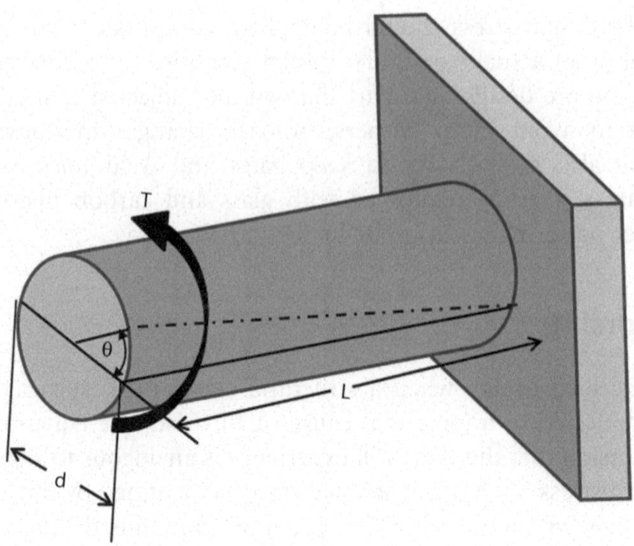


**Figure 6.12** Solid round shaft restrained at one end.

where $T$ is the twisting moment, $L$ is the length of the shaft, $J$ is the polar moment of inertia and $G$ is the shear modulus of the shaft material. We can approximate the shear modulus using Equation 6.15 in which $E$ is tensile modulus (psi) and $\upsilon$ is Poisson's ratio. When a material is in tension and experiences a longitudinal strain, it will also undergo a lateral strain perpendicular to the longitudinal axis. The ratio of the lateral strain to the axial strain is Poisson's Ratio.

$$G = \frac{E}{2(1+\upsilon)} \tag{6.15}$$

It is a constant for any given elastic solid material [26]. For most plastics, this ratio generally runs from 0.2 to 0.4.

Referring to Figure 6.10, if we have a general purpose (GP) ABS shaft 6″ long and ½″ in diameter that will see a maximum torque of 8 in-lbs, we can now calculate how much the shaft will twist. First, using Equation 6.13 we calculate the polar moment of inertia $J$:

$$J = \frac{\pi d^4}{32} = \frac{\pi(0.5)^4}{32} = 0.00613 \tag{6.16}$$

We can calculate $G$ using Equation 6.15, the value for $E$ for ABS of 300,000 psi from Figure 6.9 and the value for Poisson's ratio for GP ABS of 0.35.

$$G = \frac{E}{2(1+\upsilon)} = \frac{300,000}{2(1+0.35)} = 111,111 \text{ psi} \tag{6.17}$$

Using $G$ and $J$ we can now use Equation 6.14 to calculate the angular displacement that the shaft will see, $\theta$:

$$\theta = \frac{TL}{JG} = \frac{8 \times 6}{(0.00613 \times 111,111)} = 0.70 \text{ radians} = 4.04° \tag{6.18}$$

Using Equation 6.12 we can also calculate the maximum shear stress $\tau$ that the shaft will see:

$$\tau = \frac{Tc}{J} = \frac{8 \times \left(\frac{0.5}{2}\right)}{0.00613} = 326.1 \text{ psi} \tag{6.19}$$

Now, using a similar shaft of 8″ long, still with a maximum torque of 8 in-lbs, assume we have determined the maximum shear stress we can allow is 250 psi and need to determine what the diameter needs to be. We can rearrange Equation 6.12 to *J/c* = *T/t*. We can solve for *J/c*:

$$\frac{J}{c} = \frac{8}{250} = 0.0320 \tag{6.20}$$

Also, substituting for *J* we find *J/c* = $\Pi c^3/2$ and rearranging we get:

$$c = \sqrt[3]{\frac{2}{\Pi}\frac{J}{c}} \tag{6.21}$$

Substituting in the value we calculated for *J/c* above we get:

$$c = \left(\frac{2}{\pi} \times 0.320\right)^{\frac{1}{3}} = 0.273 \text{ and the diameter of the shaft} = 2c = 0.546\text{”} \tag{6.22}$$

which is the required diameter to keep the stress below the allowable limit.

## 6.5 Impact

One of the most difficult properties to design for is that of impact strength or impact resistance. There are two parts to this problem, that of the design itself and that of material selection. While we know a lot qualitatively about how to design a part to maximize resistance to high speed impacts, there is no system of engineering analysis that relates all the variables of type of impact, part geometry, manufacturing variables and material to impact failure. This is largely due to the complex nature of impact failure which is affected greatly by stresses in the part, the viscoelastic nature of plastics and variations in the nature of the impact itself.

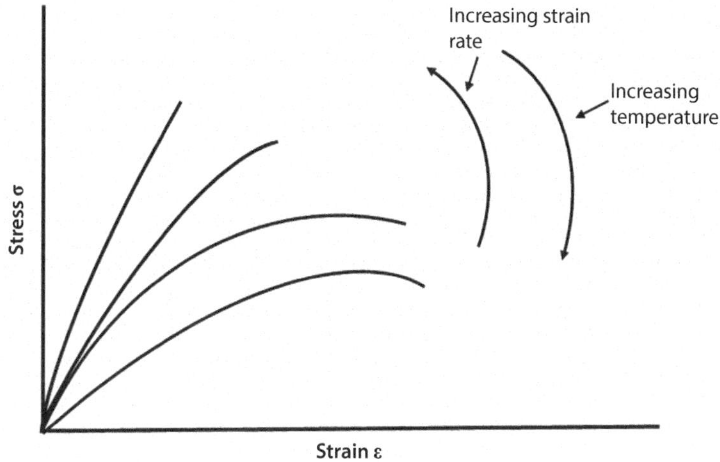


**Figure 6.13** Effect of temperature and strain rate.

As the temperature increases, materials will become more ductile. As we discussed earlier, their yield strength will decrease, but the amount of strain at failure will increase. At elevated temperatures the part will be less rigid, but it may absorb impact better because it can absorb more energy before failure. At increasing strain rates materials will fracture earlier than at lower strain rates. Figure 6.13 illustrates these effects.

Although an impact failure appears to happen suddenly, there are several things going on within the failure mechanism. First, due to the plastic's viscoelasticity, there will be elastic stretching response as the part undergoes compression and flexure as it absorbs the energy of the impact. This will be followed quickly by crack initiation and propagation and, finally, part failure as the crack propagation weakens the material to the point that it can no longer withstand the force of impact.

It is difficult to incorporate all the possible factors involved in an impact event into a standardized test and actual impacts also involve a number of nonmaterial-related factors. As a result, interpreting impact test data must be done with great care and is at best used for material selection, but even this must be done carefully. The absolute best way, of course, would be to test a prototype with the type of impacts it will see, including size of force, strain rate and temperature.

The following is a summary of some common impact tests that are used. Perhaps the most common is the Izod Impact (ASTM D256).

This is the impact test most commonly referenced in material data sheets. In this test a weighted pendulum strikes a test bar clamped at one end as a cantilever and struck near the top, applying a rapid bending stress. The impact is measured as energy absorbed in ft lbs/inch. In many ductile plastics, this test will not produce a failure so the test also allows notching the test bar to create a stress concentration that will facilitate failure. The notch geometry is carefully specified by the test but the notch geometry will have a significant effect on the test results, so application to a part design can be problematic. Also, within the ASTM D256 standard is a method for laying the test bar horizontally and supporting it on both ends. This is called the Charpy test and can be done with both a notched and unnotched sample. When comparing Izod or Charpy data it is important to be sure that the test samples and conditions are comparable. Figure 6.14 shows a sketch of the sample bar configuration used for this test. The standard allows for the width of the bars to be between 3.0 mm to 12.7 mm. The radius specified for the notch on the notched bar is very critical to the test results.

The Drop Weight Test (ASTM D3029) is probably the best of current tests to use for general material characterization. It measures the energy required to initiate fracture and is easy to run. Speed of loading and temperatures can be easily varied [27]. The test sample is a disk that is supported by a ring on its periphery and a weight with a radiused nosepiece is dropped on the center of the disk. In the procedure for a constant height test a series of impacts are made (on fresh samples) with increasing weights until failure is observed. From this data a mean failure (energy required to produce 50% failures) can statistically be calculated. This test is very similar to ASTM D2444, Impact Strength of Pipe, except that the samples are marked and rotated to be sure that some of the tests are in line with the weld lines. The drop weight test

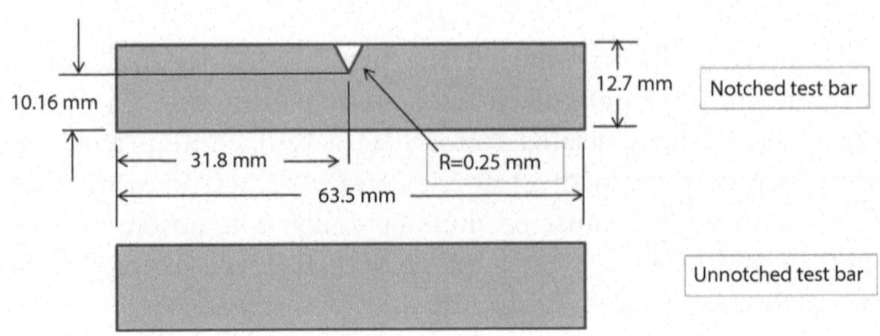


**Figure 6.14** Impact test bars.

does require a fair number of samples and does not work well with flexible or very high modulus materials.

## 6.6 Other Elevated Temperature Considerations

In addition to the mechanical properties discussed previously, the designer must consider the effects that elevated temperatures will have on other properties that will be important to the design function of the part being designed. Choosing the right material for an application at elevated temperatures can be complex because most properties will be affected but each will vary in the amount it is affected and the degree each is affected will vary for each different plastic material. For example, one plastic resistance to a particular chemical may be substantially reduced as the temperature increases, while a different plastic may see little change with temperature increases to the same chemical. Chemical resistance tables will list the temperatures that the materials were tested at. Never assume that because a material is resistant to a chemical at lower temperatures that it will also be resistant at higher temperatures. If no data is available, tests can generally be run in about 30 days that will confirm suitability for service.

As part of the design process, all of the important properties to function need to be identified, the operating temperature range identified, and each important property needs to be vetted for performance in that temperature range.

Dimensional stability can be affected in several ways by elevated temperatures. Material coefficients of expansion are not always linear and it is important to look at the specific temperature range when designing a tight toleranced part. As a fabricated plastics part is in service at an elevated temperature, there will be a relaxation of internal stresses built in the fabrication process which can lead to some additional shrinkage and warpage. This is especially true in higher pressure processes like injection molding.

Semicrystalline plastics will develop a variation in degree of crystallinity depending on process conditions such as melt temperature, mold temperature, cooling time and cooling rate. When a semicrystalline part is subjected to elevated temperatures additional crystallization can occur which can cause some additional shrinkage [28]. The degree of crystallization can also significantly affect other physical properties. If this is an issue, a post-processing annealing process can be considered.

The UL temperature index test, also called the relative thermal index (RTI), is a good tool for evaluating the suitability of a particular plastic and of specific properties at elevated temperatures. The RTI measures a plastic material's ability to retain a particular physical property above a certain level after exposure to elevated temperatures and/or extended periods of time. The testing is conducted in accordance with Underwriters Laboratories, Inc., Standard for Polymeric Materials – Long Term Property Evaluations (UL746B). The RTI can be thought of as the maximum temperature below which the material will retain sufficient properties (typically 50% of the original value) for the intended application over a reasonable product lifetime. More than one Relative Thermal Index may be appropriate for a given material depending on the property requirements for a given application. RTIs are available for selected electrical insulative properties, mechanical impact properties and mechanical strength properties. RTIs are available on samples of varying thicknesses. Normally a thick sample will age more slowly than a thin sample. The specimen thickness is noted in the data results. You should select the data from a sample with a thickness as close to the design thickness as possible.

Weathering for an extended period can also have negative effects on plastic properties due to heat from the sun. UV damage and oxidation, brittleness, loss of impact, loss of tensile strength and yellowing may occur. If weathering might be an issue, a possible solution would be to protect the exposed surface with a paint or UV-resistant coating.

In these types of applications the making of prototypes and testing is the best path whenever that is possible.

# 7

# Enclosures

Enclosures are used for many purposes in product designs, including providing product aesthetics, providing structural support, protecting the product they contain from impact and contamination, providing electrical isolation, and providing a safety barrier.

When designing a product that does something, such as a printer, hand tool or copying machine, besides the obvious safety and performance reasons, we want something to hide the inner workings from the consumer and make the product look visually pleasing. We do this with an enclosure or cover. Unfortunately, this is often the last design component considered and it must be designed to fit around other design components, usually within a fairly restricted design envelope. We want the cover to take up the minimum amount of space to keep the overall envelope of the product design as small as possible and provide a pleasing look while protecting the product from the external environment and protecting people from the product.

## 7.1 Cosmetics

The ability to achieve the type of appearance specified will be highly dependent upon the product design. For a high gloss, shiny appearance, the tooling that produces the cover will have to be made of a good quality material and be highly polished. Also, the shinier the surface, the more likely you can see a "shadow" from ribs and bosses on the back surface, particularly when using lighter colored materials. Care has to be taken to minimize these features. Gates need to be in the thickest section of the part and as large as possible. Parting lines and ejector pins need to be adequately vented to ensure the easiest flow possible into the mold. Ribs and bosses should be designed using the good practices outlined in Chapter 5. A medium or low gloss is better at hiding minor defects if that finish will be adequate for the intended use. Matte or satin finishes are excellent at hiding minor imperfections, but they can be problematic to match if an engineering change or a repair is made. Textured surfaces are common and are very good at hiding imperfections and also when you are trying to match covers made from different materials or produced by different suppliers. When textures are used, sometimes additional draft will be required to allow for release of the texture detail without marring the part surface. Textures also have the problem of being difficult to match if an engineering change or a repair is made. When matching covers of different plastic materials or plastic and metals, sometimes the only recourse is to paint them, particularly when a class "A" match is required.

On larger covers, gating of the part usually cannot be done effectively from only the edge and must be located somewhere within the surface of the cover. The gate placement can affect part shrinkage, location of weld lines, molding cycle and part quality. Generally you should gate in the thickest section if possible. Sometimes multiple gates are required to give adequate flow, but complicate shrinkage, weld lines and cost. Gates on the face of a cover can be hidden by covering them with a label or placing them behind a smaller cover or device that fits into the cover. Gates can be machined if required. This, of course, adds cost, but sometimes is a reasonable solution. Valve gating is a technique where the plastic is introduced through a hot bushing and orifice into the mold and a hydraulically actuated pin in the bushing shuts off in the orifice flush with the surface of the part. The finished gate is flush with the part surface and all that is usually visible is the appearance of a ring. This has the additional benefit of allowing fairly large gates, which help the flow into the part. Valve gating systems are very commonly used and the only disadvantages are a higher tooling cost and a little more process difficulty to start up versus a standard cold runner system.

Sometimes the cosmetic requirements are such that absolutely no gating or marks are allowable on the outside surface of the cover. In these cases all gating, ejector pins and markings will have to be on the backside of the cover. This can be accomplished by turning the part around to inject the plastic in the backside and building an ejector system in the stationary mold half (called a reverse ejection mold) of the mold. This will accomplish meeting the cosmetic requirements but will significantly add to the tooling costs.

Normally, parts will need an ejector system to assist removal of the part from the tool. Ejectors, either mechanical or air ejectors, will leave a mark on parts so they should be located on non-visual or nonfunctional surfaces when possible. When designing ejectors, they need to have enough surface area to prevent "pushing" into the part during ejection. This would result in a dimpling appearance on the side opposite the ejectors. If desired, the ejectors can be disguised or hidden by using stripper plates, rings and decorative embossings.

Around the part periphery where the tool opens there will be a parting line left on the part, usually about 0.0005″ to 0.002″ (0.013 mm - 0.05 mm) high. Also, over time a parting line will tend to get worse. This can't be made to go away, but with careful planning in the part and tool design phase it can be minimized and/or hidden. Flat, straight parting lines are easier to maintain than compound or curved parting lines. By carefully selecting where the parting line is located, it can be completely hidden from view. Figure 7.1 shows a very good technique for this by locating the parting line on a corner, as is shown by the part on the right. It will still be there but it will not be noticeable. On the part on the left, there will be a faint line around the part at the intersection of the radius with the flat

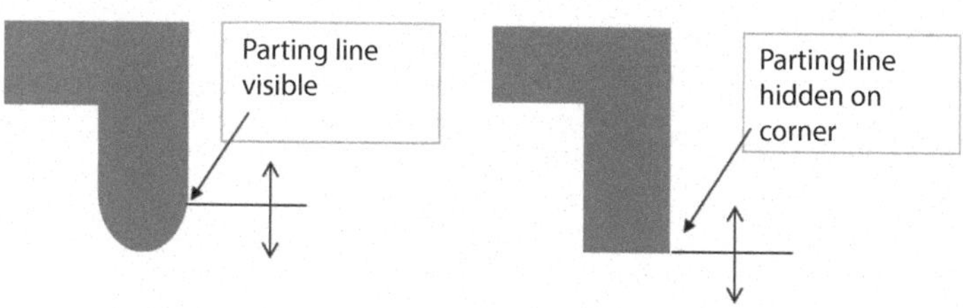


**Figure 7.1** Parting line location.

surface. With a textured surface, the parting line may be even more noticeable as it will visually break up the pattern of the texture.

Most covers require information on them such as recycle codes, flammability ratings, manufacturer, and date codes. The least expensive way to do this is to mold them into the cover, but care must be taken on how the engraving is designed and where they are located. Raised lettering (lettering engraved into the tool, raised on the plastic part) is the easiest and most economical, but it will take up some space, which might be an issue in very tight fits. Lettering needs to be large enough to be legible, but small enough to minimize any sinks or distortion on the reverse side. Lettering that is sunk into the plastic part costs more in tooling. It is good practice to put this information on an insert in the tool so that as the information changes, a new insert can be made and put into the tool. On very thin covers (<2.5mm, <0.100″) these inserts can cause a "shadow" on the reverse side of the cover. This can be caused by a combination of the force of pulling the engraving out of the tool, the translucence of the plastic material and the difference in cooling between the insert and the tool itself. The best way to deal with this is to locate the engraving opposite a noncritical area.

## 7.2 Structural Support

When designing an enclosure enough strength and rigidity must be provided to protect the product inside while maintaining a good external appearance and remaining within the available design envelope, and doing this at the lowest cost possible.

When designing a cover, the strength and rigidity can be increased just by making it thicker but this can have a fairly significant cost impact. When manufacturing a cover, the material cost will typically be 50–75% the cost of the cover. Also, thicker covers will increase the processing time. Doubling the thickness of a cover will increase the cost of the cover in excess of double. We can more effectively increase the cover's stiffness by incorporating reinforcing ribs into the design.

From Chapter 6 we have the equation for the flexural formula for beams, which is reprinted below:

$$S = \frac{M_c}{I} \tag{6.1}$$

From a mechanical design standpoint a cover is basically a beam. If we can increase the moment of inertia of the cover, *I*, we will increase its ability to resist a bending moment, *M*, and reduce the average stress *S*.

Let's look at a simple cover, 4″ across and 1/4″ thick. From the formula in Figure 6.5 we can calculate the moment of inertia for the cover, $I = bd^3/12 = (4 \times .25)^3/12 = 0.00521\ \text{in}^4$. If we increase the thickness of the cover by 1/16″ to 5/16″ and we recalculate the moment of inertia we find that the additional thickness has increased the moment of inertia to $0.0107\ \text{in}^4$. Adding a 25% increase in thickness has increased the moment of inertia by 95%. But remember that increasing weight also increases cost, so we also want to be sure that the increase is more than proportional to the weigh increase. The weight increased by 25%. We can divide the increase in *I* by the increase in weight and find that we increased I/w by 56%. This is a good metric to use when comparing the value of several different designs. Table 7.1 shows these numbers for several increases in thickness.

We can achieve significant increases in the moment of inertia by changing the geometry of the design instead of increasing the thickness and can get the increase in I that we need without adding the cost that goes with an increase in weight. The simplest way to do this is to incorporate a T beam, or a rib, into the cover. We will start with our 4″ by 1/4″ cover above and incorporate a 1/4″ high by 1/8″ thick rib into the design. In Appendix B we can find the formulas we need for this rib structure, shown in Figure 7.2.

We first need to calculate the neutral axis *c*:

$$c = d - \frac{(d^2t + s^2(b - t))}{2(bs + ht)} \tag{7.1}$$

or using Microsoft Excel® formula

**Table 7.1** Effect of wall thickness on moment of inertia.

| Cover 1 × t | Moment of Inertia I | % increase in I | I/Weight |
|---|---|---|---|
| 4″ × 1/4″ | 0.00521 | 1 | 1 |
| 4″ × 5/16″ | 0.01017 | 95% | 76% |
| 4″ × 3/8″ | 0.01758 | 238% | 158% |
| 4″ × 1/2″ | 0.04167 | 800% | 400% |

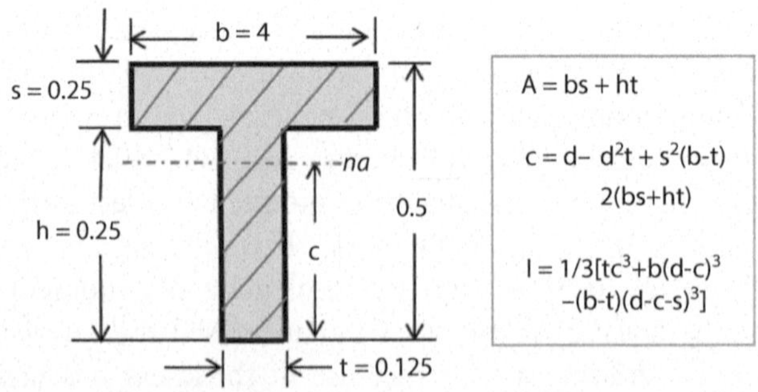


**Figure 7.2** Moment of inertia , "T" rib.

$$c = d - ((d \wedge 2^*t) + (s \wedge 2^*(b-t)))/(2^*((b^*t) = (h^*t))) \quad (7.2)$$

we find c = 0.3674. Substituting this along with the values in Figure 7.3 into the equation for the moment of inertia

$$I = \left[tc^3 + b(d-c)^3(b-t)(d-c-s)^3\right]/3 \quad (7.3)$$

or using Microsoft Excel® formula

$$I = (t^*c \wedge 3 + b^*(d-c) \wedge 3 - (b-t)^*(d-c-s) \wedge 3)/3 \quad (7.4)$$

we find that $I$ = 0.0073. By adding the rib to the structure we have increased the moment of inertia by 39%, from 0.00521 to 0.0073. Repeating the calculations for several other rib dimensions we can create a comparison, as shown in Table 7.2.

Several items of interest can be observed. First, increasing the rib height will have a much greater effect than increasing the width; and also the difference between the increase in I and the ratio of change in I to change in weight is much less than the example above where we just increased the wall thickness. Also, comparing Table 7.2 to 7.1 we find that adding a tall rib can be as effective as increasing wall thicknesses, until you make an extreme change such as doubling the wall.

When adding ribs to an enclosure you can sometimes cause a problem known as "oil canning." When this happens the cover will become convex or concave to the plane of the cover and will not lay flat. Visually, this does not look good, and if you have several covers side by side, it can ruin the aesthetic lines of the device they are covering. This happens as the heavier sections through the ribs shrink more than the

**Table 7.2** One quarter inch (1/4″) thick cover with ribs added.

| Rib Dimensions | Moment of Inertia I | % Increase in I | I/Weight |
|---|---|---|---|
| 1/4″ × 1/8″ w | 0.00727 | 39% | 38% |
| 1/4″ × 1/4″ w | 0.00921 | 77% | 72% |
| 1/2″ × 1/8″ w | 0.01478 | 184% | 173% |
| 1/2″ × 1/4″ w | 0.02344 | 350% | 311% |

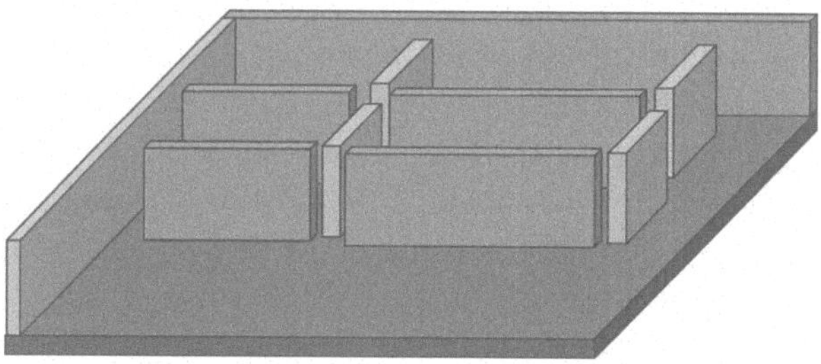

**Figure 7.3** Segmented rib design.

cover surface. The additional stress causes the cover to buckle. This problem is worse with very thin covers. A design solution to this can be achieved by using segmented ribs, as shown in Figure 7.3.

By providing a discontinuity in the ribs across the back of the cover, the effect of the shrinkage on the cover surface is reduced with only a minor reduction in the part stiffness.

When you are adding ribs to the backside of a cover for structural support you need to adhere to the 60% rule described in Chapter 5 to avoid cosmetic issues on the backside of the cover. The ribs should also be oriented along the axis of bending to provide maximum stiffness.

In some situations, because you have restricted space that limits the height of your ribs, or on a very large cover, you cannot follow this design guideline to get the structural and cosmetic results that are required. In these situations a very effective solution can be using gas assist injection molding that was discussed in Chapter 4. Gas assist injection molding is a process that produces large plastic parts with complex designs and superb cosmetic and surface finishes and allows the incorporation of very thick ribs without leaving sinks on the front (cosmetic) surface.

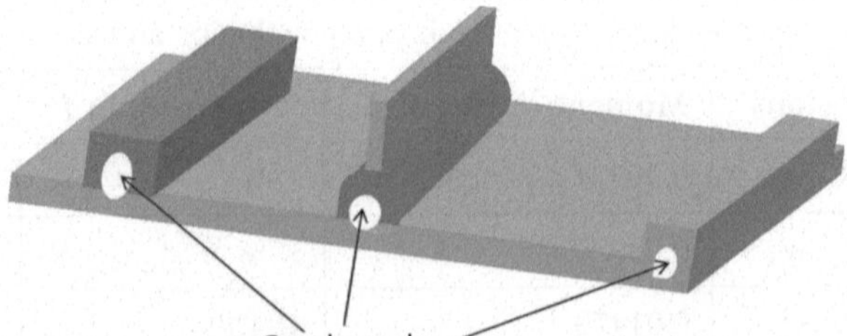


**Figure 7.4** Examples of gas assist channels.

With gas assist, the ribs have to be distinctly larger than the adjoining surface, minimally 1.5 times the wall thickness to up to as much as three times the wall thickness, in order to provide a soft plastic melt channel for the gas to pass through. You want the gas channels in the direction of material flow as much as possible and closed circuit (intersecting rib) channels should be avoided. The ribs should extend as close as possible to the last areas of the part to fill. Any corners or fillets on the ribs should be generously radiused. Remember, you do not have the concerns with sink here that you have in conventional molding.

The pressurized gas pushes the molten resin tight against the cavity walls until the part solidifies, and the constant, evenly transmitted gas pressure keeps the part from shrinking. Figure 7.4 shows a cover and some sample ribs with gas channels that would be developed in a gas assist molding process.

In addition to providing a thicker rib section with no sinks, the process requires much lower pressures than conventional injection molding, which produces lower stresses and flatter parts. The lower pressures can sometimes allow the use of lower cost aluminum tools, which can further improve flatness through better thermal conductivity to cool the part. The downside to this approach is the process is a little more complex and there may be a little trial and error involved in getting the gas channels just right when the tool is developed.

Another good technique to add structural support to a cover is to use a corrugated design as shown in Figure 7.5. A corrugated design is very effective in providing additional stiffness because it moves material away from the neutral axis and increases the moment of inertia of the section.

Corrugations increase the stiffness without the shadowing problems that can occur with ribs. The trick for the designer is how to make the corrugations look like styling features.

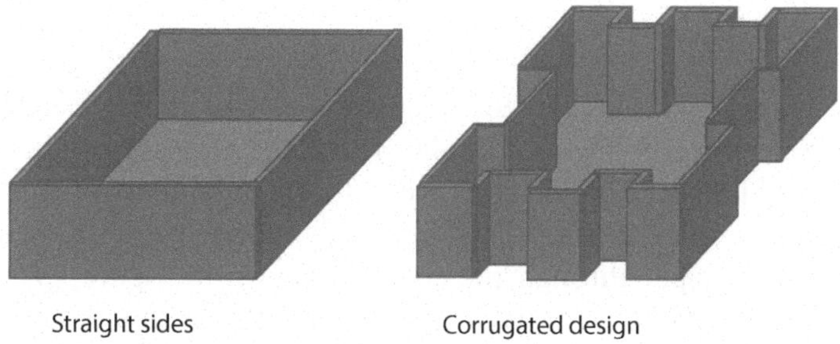


**Figure 7.5** Corrugated design for structural support.

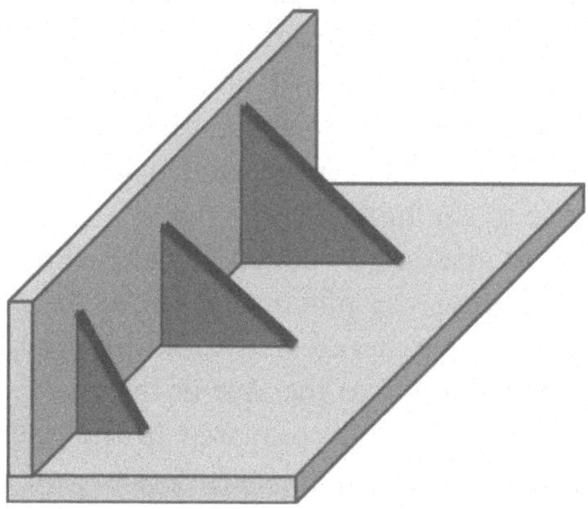

**Figure 7.6** Typical gusset configuration.

Another great technique to use if corrugations don't work is to add a slight curvature to the side walls. This will add a little stiffness and also prevent the sides pulling inward from warpage that can occur with a straight wall part. This will also work on the surface of the cover if aesthetics allow, although it will increase the cost of the mold.

Another localized stiffening feature that can be added to a part is a gusset. Gussets can be effective at reducing deflection in a specific area such as a long unsupported side wall or internal boss. They distribute a load or transfer the load from one section to another. Figure 7.6 shows a typical gusset configuration. When designing gussets you need to stick to the 60% rule for ribs and use radii to blend the top and bottom edge of the feature into the adjoining walls.

## 7.3 Ventilation

Frequently an enclosure will be housing a device that generates heat and there needs to be features designed into the enclosure to permit air flow for ventilation. This is done with the incorporation of open slots or louvers into the enclosure. This can easily be done with open molded slots that go straight through the cover as shown in Figure 7.7. This is the least expensive way of incorporating louvers within an enclosure and can be made in a simple open and shut mold.

Open louvers provide for good airflow but they also allow a view of the inner workings of the device being covered. Sometimes it is desirable to hide the device innards from view. This can be accomplished with a closed louver design as shown in Figure 7.8. This design hides the device insides from view but making these requires either cammed lifters, as shown, or by side action mechanisms, both of which will add considerable expense to the mold.

A good solution to making a hidden louver is shown in Figure 7.9 where the louver openings are put on an angle, which allows them to be molded with a simple open and shut mold by positioning the louver on a parting line angled at 45° to the face of the mold. The angled parting line will add to the tool cost compared to the open and shut open louver design but will cost less than the lifters or cams in the standard closed design and will be an easier tool to operate.

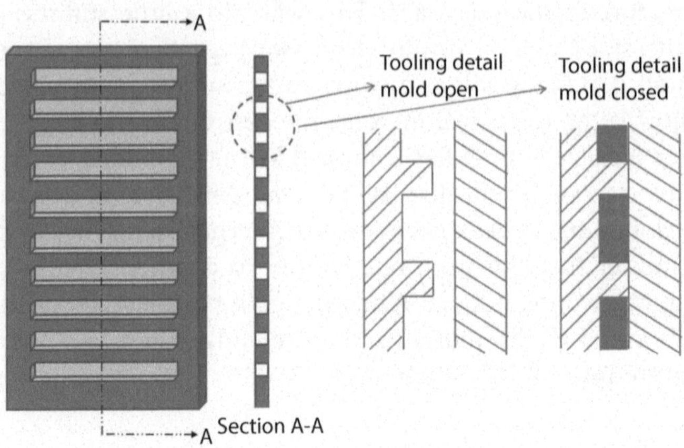


**Figure 7.7** Open louver with tool design detail.

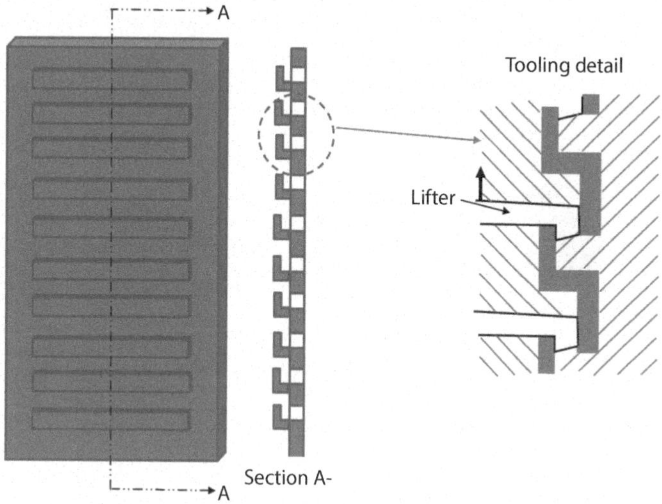


**Figure 7.8** Closed louver – Standard design.

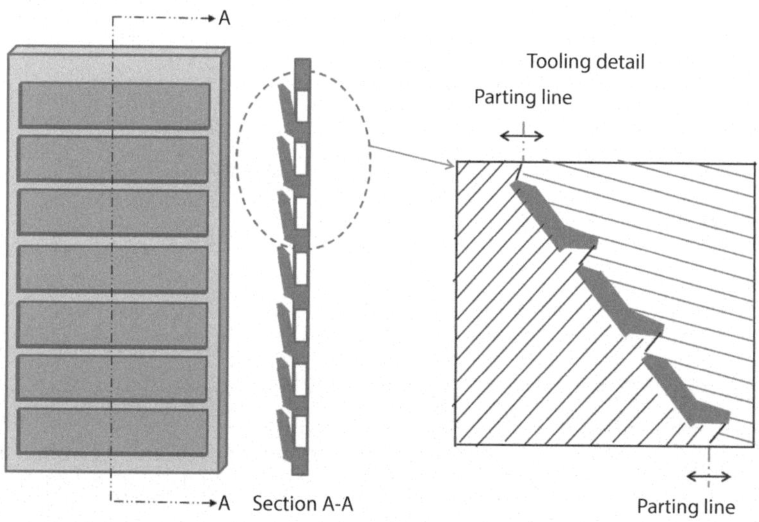


**Figure 7.9** Closed louver – Press motion design.

## 7.4 Flammability

When designing an enclosure, one of the key design requirements usually is resistance of the enclosure to burning. Most enclosures for tools, appliances and business machines have flammability requirements based

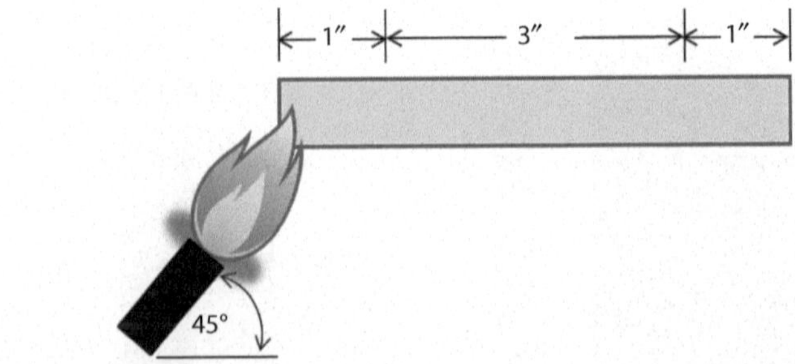


**Figure 7.10A** Horizontal burning test.

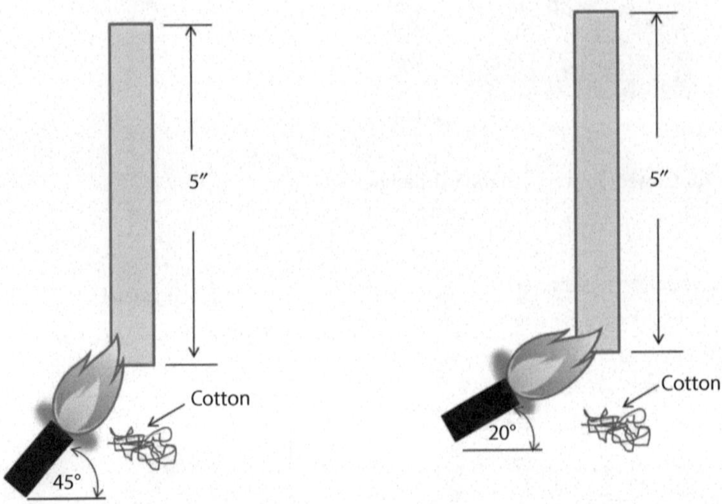


**Figure 7.10B** Vertical burning test.

on UL 94 Standard for Flammability of Plastics Materials for Parts in Devices and Appliances. Other standards like CSA and Mil standards might also apply. The UL 94 test is conducted in several different configurations. The horizontal burning test shown in Figure 7.10A holds the test specimen in a horizontal position and applies the flame to one end. The flame is positioned at a 45° angle to the test piece. The flame is applied one time and the burning rate is recorded. If the sample passes the criteria as outlined in Table 7.3, the material is awarded a UL H-B rating.

There are also two tests run with the test sample held in the vertical position. In Figure 7.10B on the left, the sample is held vertically and again the flame is applied at a 45° angle to the sample. In this test the flame is applied two separate times and combustion times are recorded.

**Table 7.3** UL 94 ratings criteria.

| Rating | Criteria |
|---|---|
| H-B | Specimens must not have a burning rate greater than 1.5 inches/minute for thicknesses between 0.120″ and 0.500″ and 3 inches/minute for thicknesses less than 0.120″.<br>Specimens must stop burning before the flame reaches the 4 inch mark. |
| V-2 | Specimens must not burn with flaming combustion for more than 30 seconds after either test flame application. Total flaming combustion time must not exceed 250 seconds for each set of 5 specimens.<br>Specimens must not burn with flaming or glowing combustion up to the specimen holding clamp.<br>Specimens can drip flaming particles that ignite the cotton.<br>No specimen can have glowing combustion remain for longer than 60 seconds after removal of the test flame. |
| V-1 | Specimens must not burn with flaming combustion for more than 30 seconds after either test flame application.<br>Total flaming combustion time must not exceed 250 seconds for each set of 5 specimens.<br>Specimens must not burn with flaming or glowing combustion up to the specimen holding clamp.<br>Specimens must not drip flaming particles that ignite the cotton.<br>No specimen can have glowing combustion remain for longer than 60 seconds after removal of the test flame. |
| V-0 | Specimens must not burn with flaming combustion for more than 10 seconds after either test flame application.<br>Total flaming combustion time must not exceed 50 seconds for each set of 5 specimens.<br>Specimens must not burn with flaming or glowing combustion up to the specimen holding clamp.<br>Specimens must not drip flaming particles that ignite the cotton.<br>No specimen can have glowing combustion remain for longer than 30 seconds after removal of the test flame. |
| 5VA | Specimens must not have any flaming or glowing combustion for more than 60 seconds after the five flame applications.<br>Specimens must not drip flaming particles that ignite the cotton.<br>Plaque specimens must not exhibit burn through (a hole). |

(*Continued*)

**Table 7.3** cont.

| Rating | Criteria |
|---|---|
| 5VB | Specimens must hot have any flaming or glowing combustion for more than 60 seconds after the five flame applications.<br>Specimens must not drip flaming particles that ignite the cotton.<br>Plaque specimens may exhibit burn through (a hole). |

Also note that cotton has been added below the flame to catch any drips or flaming particles. Depending on the performance of the sample a rating of V-2, V-1 or V-0 can be awarded.

On the right-hand side of Figure 7.10B is the Vertical Flame 5V test, whose requirements are the most stringent of all. The flame angle has been adjusted to 20° and the flame is applied 5 times. The flame must self-extinguish and no burning particles can fall on the cotton. This test can also be run with a plaque instead of a test bar. Passing of this test will give either a 5VA or a 5VB rating. Table 7.3 shows all the ratings and the performance that is required for a material to earn each one.

Materials that pass this test are given a "UL Yellow Card" that specifies the material, what thickness(es) was tested, and the rating achieved. If UL 746 RTI testing was also done, these results as well will be shown on the UL Yellow card. This data is available from the material suppliers as well as from a number of material data bases.

When you are designing enclosures, different thicknesses of materials will have different ratings, so if your requirements are V-0, a material that meets that for a 1/4″ (6 mm) cover might not meet it for a 0.100″ (2.5 mm) cover. Foam plastics will have a separate rating from solid plastic of the same material. In a complex piece of equipment, for example, a copy machine, different parts of the machine may have different flammability requirements. Generally speaking, plastics with the higher flammability ratings will cost more.

## 7.5 Electrical Considerations

Plastics are generally thought of as excellent insulators, which they are, but can deliver a broad array of electrical properties to the designer. When designing a cover or enclosure, depending on the type of cover and the location, a material that is insulative, antistatic, static dissipative, or EMI shielding may be required. Figure 7.11 shows the volume

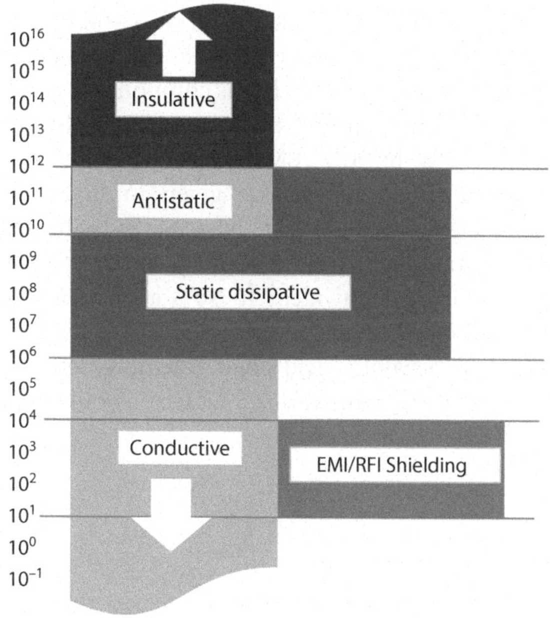


**Figure 7.11** Volume resistivity of different material types.

resistivity levels that define each of these. Plastics materials can perform all of these functions.

Electromagnetic interference (EMI) and radio frequency interference (RFI) are particularly important concerns for covers of computers, printers, communication equipment, electrical tools, and appliances. Electromagnetic and RFI emissions can affect the performance of these items and they can generate emissions that will affect other devices. EMC is defined as the ability of an electronic device to function normally without interference from or interfering with another electrical device. Shielding these devices can be provided by a conductive barrier which will transmit the electrical interference to a ground. This is called a "Faraday cage" after Michael Faraday who developed this concept. Most plastics are excellent insulators and will not function as a conductive barrier. In these cases the choices are to coat the cover with a conductive coating, such as a conductive paint or a vacuum metalized coating, or to add a conductive shield between the covers and the internal electronics. A better alternative might be to use one of a number of plastic materials that are specifically formulated with metal fillers and other materials to enhance conductivity.

Electromagnetic energy has both an electrical component and a magnetic component that move in a wave called electromagnetic radiation. EMI shielding must protect from both components. Electromagnetic energy "attenuation" is the amount that the intensity of an electromagnetic signal is reduced by the EMI shielding and is measured in decibels (dB). The formula for this is given in Equation 7.5, where *Efe* = the electric field strength effectiveness, $E_i$ = the incident field strength in volts/m and $E_t$ = the transmitted electrical field strength in volts/m. A 20 dB rating is equivalent to 90% attenuation, which would yield minimal shielding, a 60 dB level would be equivalent to a 99.9% attenuation level, which would yield good shielding, and a 100 dB level would give 99.999% attenuation, which would be excellent shielding [29].

$$Efe = \frac{20 \text{ Log } E_i}{E_t} \tag{7.5}$$

The seam where covers come together or where the cover meets the frame can be a potential source of EMI leakage and can reduce the shielding effectiveness. These gaps in the seams are called "slot antennas." The effect can be minimized or reduced by incorporating an overlap joint, a "V" groove joint or a tongue-in-groove joint at the seam. A very effective, but more expensive approach, would be to add a conductive gasket made up of a metal-filled flexible polymer or a wire mesh.

Table 7.4 gives some examples of plastic compounds that will work in EMI/RFI shielding.

When designing a cover that you want to function as an electrical insulator, such as a cover for a small power tool or appliance, the cover becomes a part of the electrical circuit it is insulating. It will introduce a power loss factor and a capacitive reactance that can affect the

**Table 7.4** Examples of EMI/RFI shielding plastic compounds.

| Material | Resistivity |
|---|---|
| ABS 5% Stainless Steel Fiber | $10^3$ Ω/cm |
| ABS 10% Stainless Steel Fiber | <10 Ω/cm |
| Polycarbonate-ABS 5% Stainless Steel Fiber | $10^3$ Ω/cm |
| Polycarbonate 5% Stainless Steel Fiber | $10^3$ Ω/cm |
| Thermoplastic Polyurethane 10% Stainless Steel Fiber | $10^3$ Ω/cm |

operation of the circuit. Volume and surface resistivity are important (you want a material with a volume resistivity of greater than $10^{12}$ Ω), as they measure current leakage across the surface of the cover and through the cover. The dielectric constant measures the contribution of the cover to capacitance and power loss in AC circuits and dissipation factor that measures the power loss dissipated as heat due to the electric fields in an AC circuit [30]. When checking the data sheets you want to be sure that the ratings are good for the service temperatures that are planned.

Lastly, a word about regulations and standards. There are a large number of standards around covers and enclosures depending on the product, product use, and geography that the product will be used in. UL746C, which we have already discussed, covers small tools and appliances. Live electrical enclosures are regulated by the National Electric Manufacturers Association (NEMA) National Electric Code (NEC) in most areas of the United States. The Canadian Electrical Code is published by Canadian Standards Association (CSA) and International Electro-technical Commission IEC 60364 is used throughout Europe. When designing an electrical enclosure, be sure to investigate the product requirements for the countries the product may be sold in.

# 8

# Packaging and Containers

Packaging is one of the largest uses of plastic materials. Almost every product that is produced is either put in a package for protection and marketing presence or the package is an integral part of the product itself. This chapter will cover a number of aspects of packaging but cannot cover the myriad of packaging applications that exist.

Packaging generally performs one or more of five principal functions, protecting the contents of the package, providing security for the contents of the package, providing information about the contents of the package, providing brand identification and presenting (marketing) the package to the customer supply chain.

## 8.1 Impact and Tear Resistance

Impact properties are very important in packaging design, as the package needs to be able to protect the contents from an impact if it is dropped or is impacted by other packages during shipment and handling. For many applications, such as a heavy item packaged in a plastic container, Izod and Charpy impact data discussed in Chapter 6 are relevant. Additionally, a few other tests will be more relevant for some packaging applications.

For thin sheets and films, the Dart Drop Impact Test (ASTM D1709) more closely represents potential impact events. This test determines the weight required to cause 50% of the film tested to fail by impact from a weighted falling dart under specified test conditions. The dart will be either 1.5″ or 2″ in diameter, with a full radiused nosepiece. Data for this test is readily available from material suppliers and some online databases. ISO Standard 7765-1 is a similar standard. Many companies or industries have their own proprietary impact tests for their packaging to try to emulate actual field conditions for the package.

In many packaging applications, the tear resistance of a plastic film or sheet is important. The test normally used for this is ASTM D1922, Propagation Tear Resistance of Plastic Film and Thin Sheeting, also known as the Elmendorf tear test. In this test a pendulum impact tester is used to measure the force required to propagate an existing slit a fixed distance to the edge of the test sample. Samples are cut from the plastic film in the machine direction and in the transverse direction. Test results are given as tearing force in gram-force or millinewtons. Sample thickness is also reported. Because there is no direct relationship between tearing force and thickness, only results for samples of the same thickness can be compared. The results of this test can be used for the specification of material and thickness for plastic film or sheet to be used in a packaging application.

## 8.2 Strength and Rigidity

Compression properties are important in packaging applications, particularly when using foamed plastics for packaging support. Compressive strength is measured by ASTM procedure D695 or ISO procedure 604. In this test a flat sample is placed between the platens of a test material and compressed. A stress/strain plot is used to develop the compressive modulus. Other strength and rigidity considerations would be the same as discussed in Chapter 7 – Enclosures.

## 8.3 Barrier Properties

One of the primary purposes of a packaging material is to keep the contents of the package inside the package and to prevent anything on the outside of the package from getting inside the package, including gases and liquids. Permeation is the process by which a chemical moves through a material on a molecular level. Knowledge about the exact permeation rates through a material is therefore essential when designing a package. We determine how easily a gas or liquid can pass through a material by the measure of a property called permeability. The lower the permeability, the more difficult it is for a gas or a liquid to pass through it. The material acts to block the passage of the gas or liquid and, therefore, permeability is called a barrier property. The permeability of the package (materials, seals, closures, etc.,) needs to be matched with package contents as well as the required shelf life and external environmental conditions.

The permeability of a plastic to liquids and gases is dependent on a number of factors. The size of the molecules of the gas or liquid is extremely important. Small molecules can work their way through the polymer structure much more easily than a large molecule can. Permeability will increase as the temperature increases because the additional energy will open up the polymer structure, making it easier for molecules to pass through. As the pressure of the air or gas increases, its permeability through a given material will increase. A polar molecule will be more permeable (will more easily pass through) to a polar plastic like nylon and a nonpolar molecule will be more permeable to a nonpolar plastic like polyethylene. A semicrystalline plastic which is tightly packed together and bound by secondary bonds will be much less permeable than a more open structured amorphous polymer.

The permeability of a plastic is measured by the diffusion coefficient or permeability coefficient. The diffusion coefficient is a measure of the permeation of a gas or liquid through a specific material. A high diffusivity coefficient is associated with high permeability. Diffusivity coefficients for different gases and liquids to plastic materials can be obtained from material suppliers or found in a variety of general databases. The data is generally given at room temperature (73 °F or 23 °C). The data can be converted for other temperatures using Equation 8.1, which is a form of the Arrhenius equation,

$$D = D_0 \, e^{-\left(\frac{A}{RT}\right)} \tag{8.1}$$

where $D$ is the diffusion coefficient at the desired temperature, $D_0$ is the diffusivity at room temperature, $A$ is the activation constant which measures the energy required for a gas or liquid to pass through the plastic material, $R$ is the ideal gas constant and $T$ is the absolute temperature measured in degrees Kelvin [31].

The diffusion rate of a material in a steady state condition where concentrations remain the same can be described mathematically by Fick's first law of diffusion given in Equation 8.2,

$$F = -D\frac{dC}{dX} \tag{8.2}$$

where $F$ equals the flux (the rate of transfer of a material per unit area), $D$ = the diffusion coefficient, $dC$ is the change in concentration from one side of the material measures to the other and $dX$ is the measure from entrance point to point of measure.

The more common case for when the diffusion rate varies over time is described mathematically by Fick's second law of diffusion, given in Equation 8.3,

$$\frac{\delta C}{\delta t} = D\frac{\delta^2 C}{\delta x^2} \tag{8.3}$$

where $C$ equals the flux concentration of the gas or liquid, $t$ is the permeation time, $D$ is the diffusion coefficient and $x$ is the distance from one edge to the point of measurement and $\delta$ are partial differentials of thickness and time [32].

Permeability data for water vapor transmission (WVTR) is usually given in grams × mils /100 $in^2$× 24 hour, where mils is thickness of the material in thousandths of an inch (0. 001″). Permeation for gases usually is given in $cm^3$× mil /24hrs/100$in^2$× atm, where atm is pressure in atmospheres. Table 8.1 gives an example of data for water vapor transmission rate for several selected generic polymers.

To apply this data you need to know the material being used, the surface area and the thickness of the design being considered and the time period that is of interest. For example, you are designing a container to hold water that is 0.0625 inches thick (62.5 mils) and has a surface area of 27 square inches. You want to know how much water the container will lose in three months. Your initial material choice is PVC. The bottle will always be stored at room temperature or below, so conservatively, you will do the calculations as though the container is stored at a constant room temperature (at any periods below room temperature, the permeation rate will be less). From Table 8.1, the room

**Table 8.1** Generic WTVR for selected plastics.

| Generic WVTR for Selected Plastics | |
|---|---|
| Material | WVTR (gxmil /100 in$^2$ $\times$ 24hrs) |
| Polypropylene | 0.3 |
| Ethylene Vinyl Acetate (EVA) | 0.4 |
| Polyethylene Terephalate (PET) | 1.3 |
| Polyvinylchloride (PVC) | 4.5 |
| Nylon (PA) | 19 |

$$\text{WVTR} = 4.5\text{g} \times \text{mil}/100\ \text{in}^2 \times 24\text{hrs}$$

$$\text{Water loss} = 4.5\text{g} \times \text{mil}/100\ \text{in}^2 \times 24\text{hrs} \left[ \frac{27\ \text{in}^2 \times 90}{62.5\ \text{mils}} \right]$$

$$\text{Water loss in 90 days} = 1.75\text{g}$$

**Figure 8.1** Calculation of WTVR – Example 1.

temperature WVTR for PVC is 4.5g $\times$ mil /100in$^2$ $\times$ 24 hrs. The calculation is shown in Figure 8.1.

That number seems a little high. Redoing the calculation for one year we find that we will lose over 7 grams of water in one year.

Assume the number desired is no more than 0.75 grams per 3 month period. Your choices would be to increase the wall thickness or change the material. It is obvious that the wall would have to be increased significantly (actually to 0.146″), and that would significantly raise the cost of the container. We would like to find out what is the WVTR we would need to meet the 0.75 grams target. We can do this by rearranging the equation in Figure 8.1 and solving for the WTVR. This is illustrated in Figure 8.2.

A WVTR of 1.929 or less will meet the requirement of a maximum loss of 0.75 grams of water in 90 days. Looking at the material data in Table 8.1 we find that PET, EVA and PP all would meet this requirement.

When specifying a material for a specific application, there can be permeation limit requirements for several different gases and liquids, for example, carbon dioxide and water for a carbonated beverage bottle. Sometimes no material will be adequate for all requirements. Materials

$$WVTR = \text{Water loss in g} \times \text{mil/days/service/surface area in}^2/100$$

$$WVTR = \left[0.75g \times 62.5\ \text{mil}/90\right]/(27\ \text{in}^2/100)$$

$$WVTR = 1.929$$

**Figure 8.2** Calculation of WTVR – Example 2.

can be chemically modified by sulfonation or fluorination processes to improve permeability. The most common solution to this problem is multilayer co-extrusion, which was briefly discussed in Chapter 4. In this process, the crosshead film die is fed by a different extruder for each material desired. The layers can consist of a variety of polar and nonpolar polymers to produce the overall permeability characteristics that are required. Two- or three-layer film is common and films with over a dozen layers have been made. There are limitations in that the polymer layers need to be compatible, that they will stick together and not separate. Many materials will stick only to themselves and one or two other materials. There are special polyolefin-based resins that will stick to a variety of other polyolefins and are used as a tie layer. A very few, like ionomer and ethylene vinyl acetate, will stick to a variety of materials and can be used as a tie layer between quite dissimilar materials. As more layers are added, the die geometry gets quite complex and the processing more difficult.

Also, thin metallic films can be coated onto certain polymer films by vacuum deposition to reduce permeability to certain gases. Most metals have lower permeabilities than plastics due to their relatively dense crystalline structures.

Multilayer material designs are also possible with blow molding, where the coextruded parison to be blown into a container is similar to multilayer film. Also, this multilayer technology has been brought to injection blow molding but due to the complexity of tooling, fewer layers are typical.

## 8.4 Packaging Processes

Packaging applications can use all of the standard plastics primary manufacturing processes, although there are some interesting twists on several of them that will be discussed. Several of the most interesting involve the thermoforming, or vacuum forming, process.

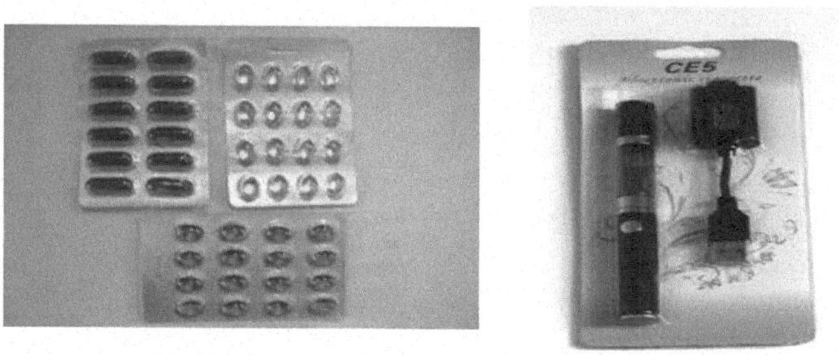

**Figure 8.3** Blister pack examples.

In a blister pack a mold is made to approximately the shape of the part/product to be packaged but just a little bit larger. A thin plastic sheet or film is formed against the mold cavity. The product is placed in the formed plastic cavity and a backing sheet is applied with an adhesive or heat. Figure 8.3 gives some examples of blister packs.

A skin pack is similar to a blister pack but in this case the product itself becomes the mold. Again the plastic film or sheet is heated and a vacuum is drawn through holes in a paper or cardboard backing, drawing a plastic "skin" around the part to be packaged.

Another permutation of the thermoforming process used in packaging is the form, fill and seal (FFS) process. In this process, a container is thermoformed and then the container is filled with the package contents such as jelly, yogurt, coffee creamer, catsup, etc. This is usually done in very high volumes and a special continuous in-line thermoforming, product fill and seal will be combined into an automatic continuous operation, as shown in Figure 8.4. The plastic film or sheet is fed off a roll and passes under a heating station. When softened, the material is formed and the cavities are then filled with product. A sealing layer is glued on top and the individual filled packages are cut out and sent to a packing line to be put into cartons or boxes. An inkjet printer can be added to provide bar codes and manufacturing or use by dates.

This operation can be further automated by replacing the roll of plastic sheet with an extruder and a sheeting die.

When designing for thermoforming applications generous radii and fillets should be used. As the plastic is formed it will be stretched and the wall will thin. In order to maintain the wall thickness, either for strength or permeability concerns, you need to compensate for this stretching. Assume you are making a thermoformed cavity 2.5″ × 2″ × 3/8″ deep. To maintain permeability, a minimum .045″ wall thickness is

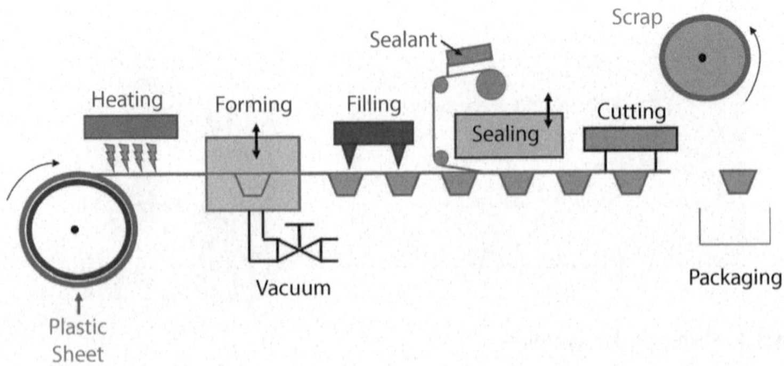


**Figure 8.4** Continuous FFS operation.

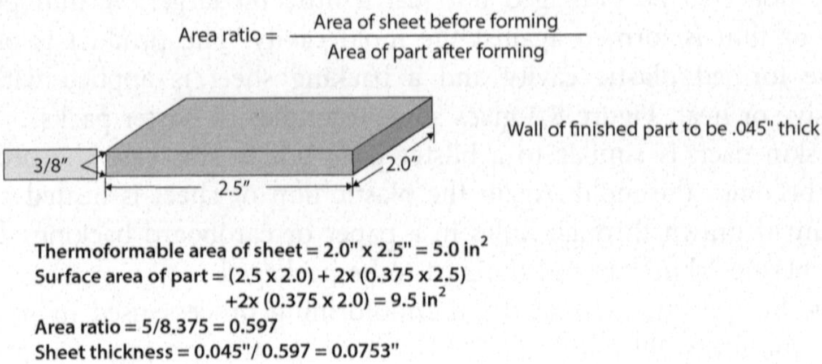


**Figure 8.5** Sheet thickness calculation.

required. Figure 8.5 shows how to calculate what the starting sheet thickness needs to be to end up with a required finished thickness.

Blow molding is used for many hollow containers and bottles. To allow for containers that have low permeability to different gases, multi-material bottles can be produced by using a co-extruder to produce a layered parison. Improvements in process control technology have greatly improved wall thickness control, which has allowed for cost saving by being able to manufacture parts closer to design minimums.

Co-injection molding also can be used to make packaging with multiple materials. For example, a closure can be made with an inner layer that is non-permeable to carbon dioxide and an outer layer that provides another attribute such as tensile strength or flexibility. In injection molding, the parts can be designed to mechanically adhere to one another so material compatibility is not as important as it is in extrusion.

Rigid molded plastic foams are commonly used for protective packaging. Normal general design practices outlined in Chapter 6 apply. Draft on part sidewalls needs to be ample, in the 3° to 5° range. Good data on mechanical properties of foams are not readily available. On critical applications prototyping and testing may be required.

A great way to protect a product from shipping and handling damage is to use a foam-in-place package. In this method, two chemical liquids, such as a di-isocyanate and a multifunctional organic alcohol, are pumped through a mixing head where a polyurethane foam is created. The product is wrapped in a polyethylene sheet or bag for protection, placed in a box and the foam is introduced. As it expands, it provides a tight, impact-resistant package, securely holding the part in position within the carton. The density of the foam can be specified to provide the combination of impact resistance and load bearing that works best for the application.

## 8.5 Printing and Decorating

When designing a package, often decorating or printing is required on the package in order to provide information about the contents, providing brand identification and presenting (marketing) the package to the customer chain. The discussion below will detail a number of methods that can be used for this.

Often it is desirable to be able to have readable information on a package such as customer names and logos, warnings, customer information and instructions. If the package is molded, incorporating this information as part of the molding is usually the most economical way to provide this information and it also has the advantage of being permanent and not likely to wear off. When using this route for the print, the lettering should be limited to about 0.010″ (0.25 mm) deep and should be at least 1-1/2 times as wide at the base as it is deep. The top of the letter should be radiused and the sides should be tapered 5°. If you are making wider letters or logos, depth is no longer a concern but the top and bottom edges should be radiused and draft should be 3–5°.

Another way to get printed information on a part is with an inkjet printer. This is an excellent choice if you need only one color (additional colors require separate printheads) and you have a lot of letters and/or different letters for each part. Using computer software, inkjet printers can print time and date stamps and bar codes. They can easily be added to almost any operation, including injection molding, extrusion, assembly and packaging. A similar technology would be laser printing, where a laser can be used to mark print or designs directly on a part or can be used to selectively burn a coating off of the part.

Hot stamping provides an effective method for creating letters, logos and patterns into a part in a wide array of colors and finishes. In this process, the information desired is engraved on a die and the die is heated and is pressed against a colored foil positioned on the surface of the part. The heat and pressure transfer the colored ink from the foil to the part. An advantage of the process is that the heat and pressure emboss the print into the part, making the information difficult to wear off. A variation of this process using a silicone rubber pad instead of an engraved die can be used to apply color to the top of raised molded-in lettering or logos.

Pad printing uses a soft rubber pad to pick up ink through a mask in the shape of the pattern or lettering desired and transfers the ink to the surface of the part. The advantage of this process is that it uses very low pressures and can print on very thin or fragile parts and it can print very easily on curved and non-flat surfaces.

Screening or silk screening is an inexpensive method to decorate flat and cylindrical plastic parts. A stencil with the information or decoration you desire is made by a photoetching or laser machining process. The stencil is placed on a screen and ink is forced through the screen and transferred to the plastic part. The screening process does not have the precision of hot stamping or pad printing but is a good method for low volumes and prototyping.

Decorations and instructions can also be printed onto a self-adhesive label and that label can be applied to the package as a secondary operation. Labels are available in transparent, opaque, metallic and embossed materials. Labels can also be strategically placed to hide a gate or cosmetic defect. Although this is one of the more costly routes, the advantage is that it is easier to change information or to apply different information to a part.

Applying decorations to the package as part of the injection molding process instead of as a secondary decorating operation can lower decorating costs as well as minimize or eliminate VOC emissions, handling, and part inventories. Application of multicolor labels and complex graphic patterns are possible. In-mold transfer decoration involves the transfer of the graphic design from a carrier film, such as polyester, to the plastic part as it is molded. The decorated film is placed in position on the mold surface to be decorated and the heat and pressure of the molding process adhere the plastic to the carrier film. The carrier film is removed from the part, leaving the graphics. In-mold decoration allows application of multiple colors and patterns in a single operation and part graphics can be updated or changed by changing images on the carrier film. The decorative film can carry sequential part numbers or bar codes. A modification of this process involves spraying a powder coating on the mold surface which is transferred to the molded part during the molding process.

# 9

# Snap Fits and Hinges

Snap fits provide one of the fastest and most cost effective methods of assembling two or more parts together and also are often one of the simplest. When designed properly, snap fits can be made to assemble and disassemble numerous times without any adverse effects on the assembly quality. Snap fits are an important design feature to use in environmentally friendly "green" designs as they are very useful in improving recyclability of a product, as they can be disassembled (if designed properly) relatively easily and the parts can easily be sorted for reuse or recycling.

Snap fits can be designed with both thermoplastic and thermoset plastics, but thermoplastics, due to their high flexibility and elongation, low coefficient of friction and resilience, which allows multiple assemblies and disassemblies, as well as their ability to be easily molded into complex geometries, make them the materials of choice for snap fits in most applications.

The unique properties of plastics to strengthen as the molecules are oriented permit manufacturing strong, flexible hinges that save parts and assembly time and are integral to the part design.

## 9.1 Snap Fit Designs

Snap fits provide a wide range of design possibilities and design solutions such as the easy attachment of one small part to another, the complex assembly of many parts together, the easy joining of dissimilar materials, caps and closures. Snap fits solve the problem of how to create assemblies that can quickly and inexpensively be assembled, disassembled and reassembled.

The most common type of snap fit joint is the simple cantilever design, as shown in Figure 9.1.

The flexible snap (catch) feature will engage in a rigid opening or cutout (detent) on the mating part as shown, or will engage with a ledge. The snap feature is essentially a simple beam of either a rectangular or more complex cross section. As the snap is engaged, the beam is deflected within the allowable strain for the material being used and as the snap feature passes the cutout or ledge, it will return to its original stress-free position, locking the two parts together. The joint may be permanent and inseparable or it can also be designed to allow the two parts to be separated for disassembly. When designing a cantilever snap the mechanical force required for the assembly operation, as well as the mechanical force required to hold the assembly together, must be considered.

There are a number of variations of the basic cantilever beam design. One of these that is quite common is the U-shaped cantilever snap, as shown in Figure 9.2. This configuration increases the flexibility of the snap's arm by increasing the effective arm length and lowers the amount of stress seen at the base of the snap. It can be used to engage a horizontal parallel surface as shown, as well as a perpendicular vertical

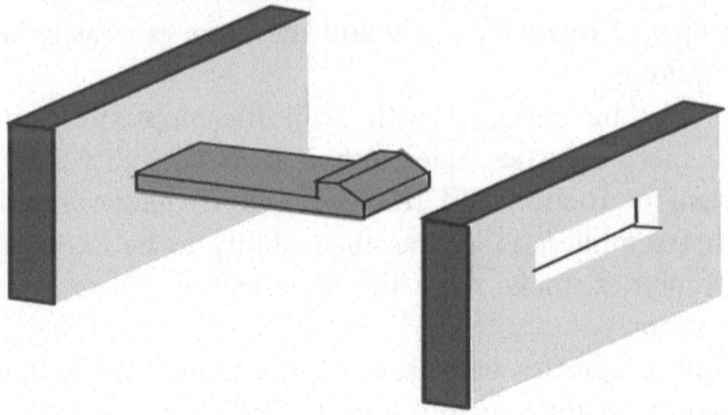

**Figure 9.1** Simple cantilever snap fit.

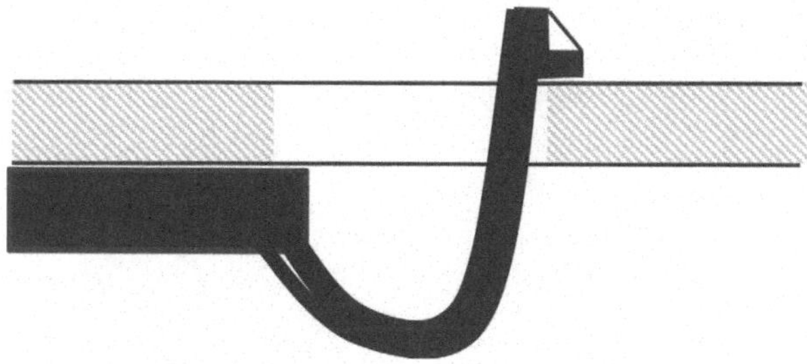

**Figure 9.2** U-shaped cantilever snap fit.

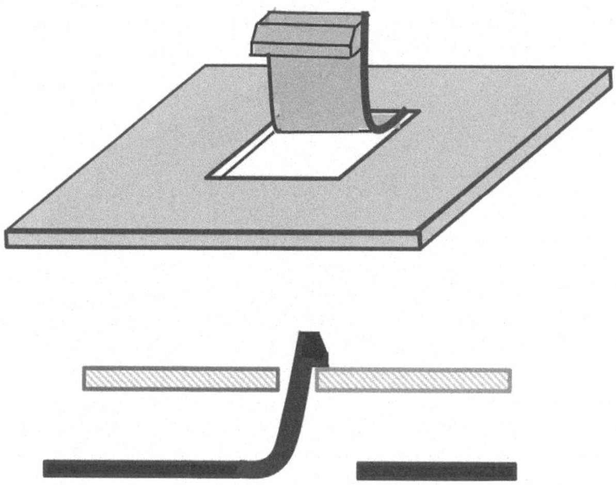

**Figure 9.3** L-shaped cantilever snap fit.

surface. The "U" shape is a little more complex than the simple beam and can have a slightly higher tooling cost to produce.

Another type that is commonly used is the L-shaped cantilever design, as shown in Figure 9.3.

Like the "U" shape, this design also provides a snap with greater flexibility, but lends itself very well to applications that have limited space available for a joining feature.

Another variant called a snap fit loop (Figure 9.4) employs a flexible detent where the catch feature now is the rigid member and the detent is the member that flexes upon engagement. This configuration will have a high resistance to accidental disengagement of the assembly. The catch feature will be an integral part of the body or frame of the mating part.

Annular snap fit joints are radially symmetrical and because of the continuous nature of the snap, relatively small undercuts can give a

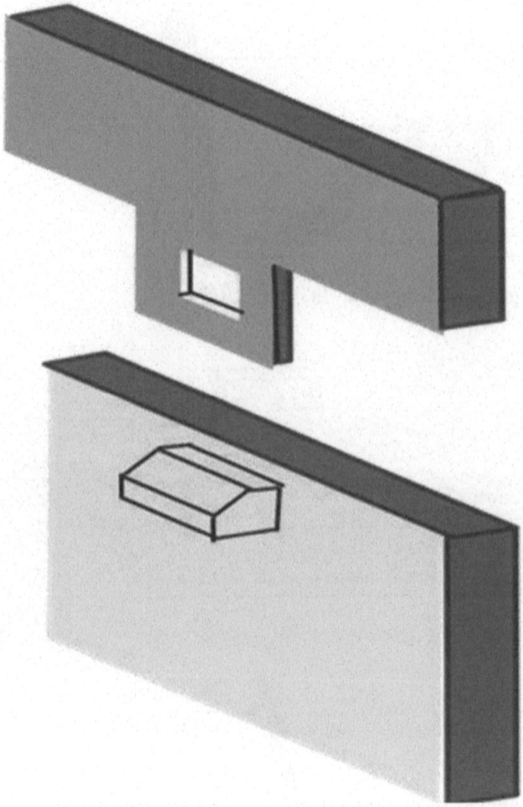

**Figure 9.4** Snap fit "loop."

fairly high pullout force. This type of design is commonly used for container closures (Figure 9.5).

## 9.2 Design of Cantilever Snaps Using Classical Beam Theory

A cantilever snap fit is structurally a simple beam and we will approach the design using many of the same principles that were used in Chapter 6, Structural Components.

A cantilever snap fit consists of a cantilever beam that has a protrusion on the end that will capture the opening or step of the mating part, as shown in Figure 9.6. The depth of this protrusion, or catch depth, will determine the amount of deflection which will be required for assembly and disassembly.

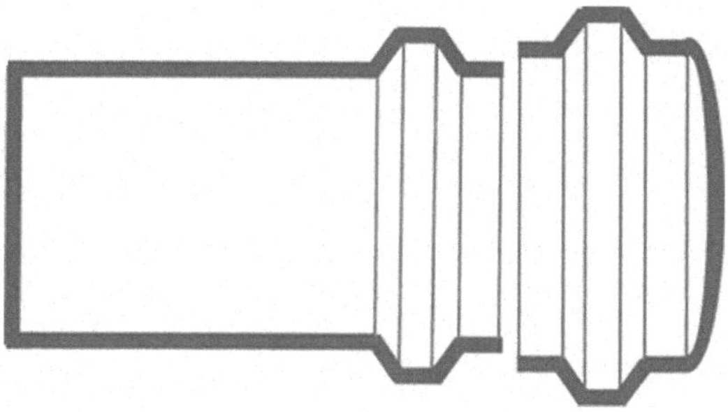

**Figure 9.5** Annular snap fits.

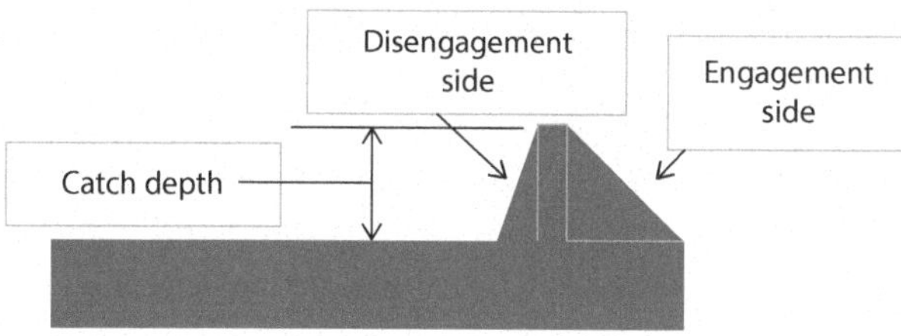


**Figure 9.6** Schematic of snap details.

The engagement side of the snap is where it pushes against the mating part during assembly, deflecting the beam downward to allow the snap to engage. This side typically will have a gentle angle to provide for ease of assembly as the two mating parts are pushed together. Typically this angle will be between 15° and 30°. A larger angle will increase the assembly force required and can make the snap very difficult to assemble. The disengagement side will usually have a much steeper angle which will make accidental disassembly difficult or impossible, depending on the intended use. Disassembly side angles are typically in the range of 30°to 50°. By modifying these angles the designer can adjust the amount of force required for assembly and disassembly of the parts. The amount of force that holds the assembly together can be increased by increasing the catch depth but this will require the beam to deflect further in order to engage the snap. As the amount of deflection increases, the amount of stress on the beam will also increase proportionally. This stress has to be kept below the yield stress of the material being used to prevent failure of the snap during assembly. The designer needs to balance the amount of catch depth desired with the beam geometry and the material modulus to produce the optimum design solution.

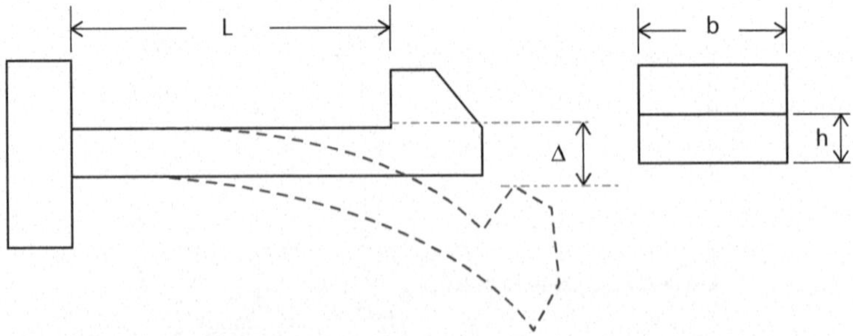


**Figure 9.7** Simple cantilever beam – Uniform rectangular cross section.

We will first look at a simple snap fit with a uniform rectangular cross section, as shown in Figure 9.7, with an arm length "L," a beam width "b" and a beam thickness "h." As the snap is engaged, it will deflect the amount $\Delta$. In this application, as the snap is engaged the beam will essentially be loaded on one end and the deflection ($\Delta$) that the beam will see is described in Equation 9.1 as:

$$\Delta = \frac{FL^3}{3EI} \tag{9.1}$$

where $F$ equals the force applied to the beam, $L$ is the length of the beam, $E$ the material modulus and $I$ the moment of inertia of the beam [33]. (This equation is slightly different than Equation 6.9 for beam deflection in Chapter 6 which had a load uniformly applied to the entire length of the beam.)

By using the equation $\sigma$ = E $\varepsilon$ from Chapter 2 and Equation 6.2 from Chapter 6, $\sigma$ = M/Z = FL/Z, and substituting, we can rewrite Equation 9.1 into a very useful form, Equation 9.2,

$$\Delta = 0.67\frac{\varepsilon L^2}{h} \tag{9.2}$$

where $\varepsilon$ equals the maximum strain on the beam. Using the strain value will ease calculations of snap fits. The strain cannot exceed the maximum allowable dynamic strain for any given material. A good rule of thumb is that most unfilled thermoplastics can withstand approximately 6% strain, while a filled thermoplastic can withstand about 1.5% strain. A more accurate method of estimating the allowable dynamic strain for any given material can be determined from the material's stress-strain

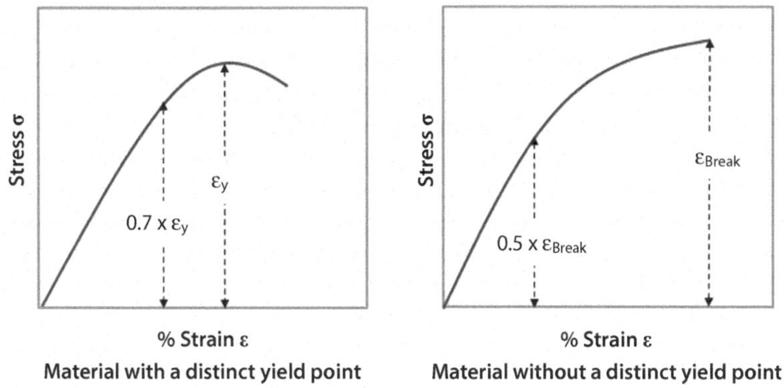


**Figure 9.8** Stress-strain curves for materials with and without yield point.

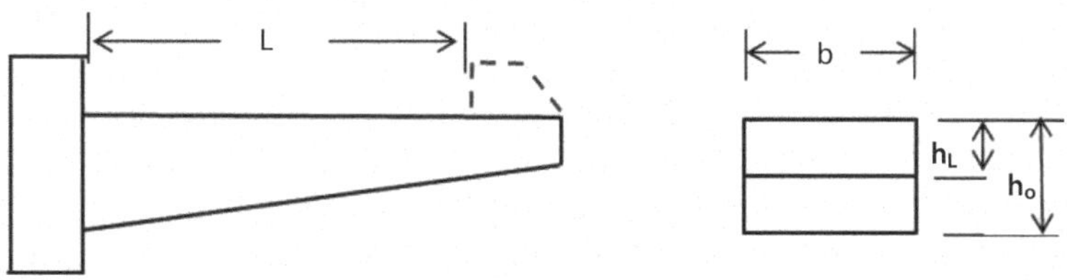


**Figure 9.9** Tapered cantilever beam.

curve. For most thermoplastic materials with a definite yield point, its allowable dynamic strain may be as high as 70% of the yield point strain. For materials that break at low elongations without yielding, like most filled thermoplastics and many thermosets, a strain limit of up to 50% of the strain at break may be used (see Figure 9.8).

These strains are based upon a single use of the snap joint. Many designs use a snap fit as a one-time permanent assembly latch. If multiple engagements and disengagements are to be used this repeated service aspect must be considered and the beam should be designed at approximately 50–60% of the recommended strain levels. In all cases, it is strongly recommended to get specific material strain level recommendations from the material manufacturer.

When snap fit arms (the cantilever beam) are designed with a uniform cross section, the stress is concentrated at the base of the beam (where it intersects with the wall of the part). A better way to design a snap fit arm is to use a tapered beam design, as in Figure 9.9, with the thickest section at the base and decreasing toward the tip section. The taper effectively decreases the beam's strain and allows for a deflection greater than that of a straight beam with the same base thickness. The taper of the beam is described by the ratio of the beam height or thickness at the small end ($h_L$) to the beam height or thickness at the large end ($h_o$). This ratio is called the proportionality constant, $K$.

Modifying Equation 9.2 for a tapered beam (setting $h = h_o$) gives us the following equations:

$$\Delta = 0.67\frac{\varepsilon L^2 K}{h_0} \tag{9.3}$$

$$\varepsilon = 1.5\frac{\Delta h_0}{L^2 K} \tag{9.4}$$

The value for $K$ for various $h_{L/}h_o$ ratios for tapered beams can be found in Table 9.1 [34].

Let's look at some examples of how to apply these equations in developing the design of a snap fit. For our first example we want to design a simple cantilever snap fit arm with a uniform rectangular cross section as depicted in Figure 9.7. The beam length, $L$, is 1.0 inches, the beam thickness, $h$, is 0.125 inches and it will be made of a generic polycarbonate material. We need to calculate the maximum deflection, $\Delta$, that can be tolerated for a *one-use* snap fit. We can solve for this by using Equation 9.2:

$$\Delta = 0.67\frac{\varepsilon L^2}{h} \tag{9.2}$$

We are given L and h and we can use Table 9.2 to find the maximum allowable strain $\varepsilon$.

From Table 9.2 Max allowable strain = 4–6%. Without specific information as to specific grade, we will conservatively use the 4% figure. Normally you would get a very specific recommendation from the material supplier for the material and grade chosen. Substituting these values we can solve the above equation as follows:

$$\Delta = 0.67\frac{(0.04)(1.00)^2}{0.125} \tag{9.5}$$

$$\Delta = 0.214''$$

**Table 9.1** Tapered beam proportionality constants (K).

| $h_L/h_0$ | K | $h_L/h_0$ | K | $h_L/h_0$ | K | $h_L/h_0$ | K |
|---|---|---|---|---|---|---|---|
| 0.33 | 2.137 | 0.50 | 1.636 | 0.67 | 1.338 | 0.85 | 1.138 |
| 0.34 | 2.098 | 0.51 | 1.614 | 0.68 | 1.324 | 0.86 | 1.128 |
| 0.35 | 2.060 | 0.52 | 1.593 | 0.69 | 1.310 | 0.87 | 1.118 |
| 0.36 | 2.024 | 0.53 | 1.573 | 0.70 | 1.297 | 0.88 | 1.109 |
| 0.37 | 1.989 | 0.54 | 1.553 | 0.71 | 1.284 | 0.89 | 1.100 |
| 0.38 | 1.956 | 0.55 | 1.534 | 0.72 | 1.272 | 0.90 | 1.010 |
| 0.39 | 1.924 | 0.56 | 1.515 | 0.73 | 1.259 | 0.91 | 1.082 |
| 0.40 | 1.893 | 0.57 | 1.497 | 0.74 | 1.247 | 0.92 | 1.073 |
| 0.41 | 1.863 | 0.58 | 1.479 | 0.75 | 1.235 | 0.93 | 1.064 |
| 0.42 | 1.834 | 0.59 | 1.462 | 0.76 | 1.223 | 0.94 | 1.056 |
| 0.43 | 1.806 | 0.60 | 1.445 | 0.77 | 1.212 | 0.95 | 1.047 |
| 0.44 | 1.780 | 0.61 | 1.429 | 0.78 | 1.201 | 0.96 | 1.039 |
| 0.45 | 1.754 | 0.62 | 1.413 | 0.79 | 1.190 | 0.97 | 1.031 |
| 0.46 | 1.729 | 0.63 | 1.399 | 0.80 | 1.179 | 0.98 | 1.023 |
| 0.47 | 1.704 | 0.64 | 1.382 | 0.81 | 1.168 | 0.99 | 1.015 |
| 0.48 | 1.681 | 0.65 | 1.367 | 0.82 | 1.158 | 0.90 | 1.008 |
| 0.49 | 1.658 | 0.66 | 1.352 | 0.83 | 1.148 | 1.00 | 1.000 |

We find that 0.214" deflection can be tolerated for this snap fit to allow engagement of the one-time snap. As the product design progresses, it is determined that the snap fit will be used multiple times. What is the design deflection that should now be used? Remember, for multiple uses, the allowable maximum strain must be reduced to no more than 50% of the maximum value. Using the 50% rule we can resolve the above equations:

**Table 9.2** Permissible strain and friction properties for selected generic materials at 73 °F (23 °C).

| **Material** | **Max Allowable Strain** | **Coefficient of Friction $\mu$** |
|---|---|---|
| ABS | 6.0–7.0% | 0.50–0.60 |
| Acetal (POM) | 7.0% | 0.20–0.35 |
| Nylon 6 | 7.0–8.0% | 0.17–0.40 |
| Polycarbonate (PC) | 4.0–6.0% | 0.25–0.40 |
| Polypropylene | 8.0–9.0% | 0.20–0.35 |
| PBT | 8.8% | 0.35–0.40 |
| 30% Glass Reinforced Acetal | 2.0% | |
| Polycarbonate/ABS Blend | 2.5% | |
| 20% Glass Reinforced PC | 1.5% | |

$$\Delta = 0.67\frac{50\%\varepsilon L^2}{h}$$

$$\Delta = 0.67\frac{(0.50)(0.04)(1.00)^2}{.125} \tag{9.6}$$

$$\Delta = 0.107''$$

As would be expected, to reduce the strain, the amount of deflection that can be tolerated will be reduced proportionately.

Next, we will design a tapered beam snap fit similar to Figure 9.9. The length of the snap fit beam L is 0.75″. The thickness at the base, $h$, is 0.375″ and the thickness at the tip, $h_L$, is 0.1875″. The undercut of the snap is 0.100″ (this is the amount the beam section must deflect, D, to engage the snap). What is the maximum strain that the beam will encounter?

To solve this, we will use Equation 9.4, which gives us strain in a tapered beam.

$$\varepsilon = 1.5 \ \frac{\Delta h_0}{L^2 K} \tag{9.4}$$

We first need to find the value for the appropriate proportionality constant K for the ratio of tip thickness to base thickness $h_L/h_0$.

$$\frac{h_L}{h_0} = 0.1875/0.375 = 0.5 \tag{9.7}$$

From Table 9.1 we find that for $h_L/h_0$ of 0.5, the proportionality constant K is equal to 1.636. We now have all the information we need to solve for $\varepsilon$.

$$\begin{gathered} \varepsilon = 1.5 \ \frac{(0.100)(0.375)}{(0.75)^2(1.636)} \\ \varepsilon = 0.0611 = 6.11\% \end{gathered} \tag{9.8}$$

In order to move the required 0.100″ the beam will undergo a strain of 6.11%.

Now assume the material of choice for this application is 30% glass-filled acetal. Looking at Table 9.2 we see the maximum allowable strain for this material is 2%. The design as is in this material will not work. The choices are to change the material or change the design. Assuming that this material is the best selection, options include changing the geometry to change the proportionality constant or changing the length of the cantilever. We can quickly look at what length would be required, as shown below. We can rearrange Equation 9.4 to solve for L. Using our 2% strain limit, we can solve for L as follows:

$$\begin{gathered} L = \sqrt{1.5 \ \frac{\Delta h_0}{\varepsilon K}} \\ L = \sqrt{1.5 \ \frac{(0.100)(0.375}{(0.02)(1.636)}} \\ L = 1.32'' \end{gathered} \tag{9.9}$$

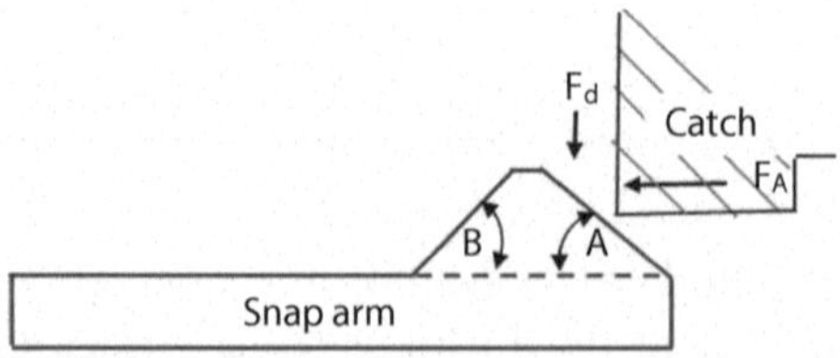


**Figure 9.10** Assembly of snap fit.

If we have room in our design envelope to increase the length of the snap arm to 1.32″, we can use 30% GF acetal and accommodate the 0.100″ movement required for the snap.

## 9.3 Assembly and Disassembly

When assembling a snap fit, as the snap is pushed into place, sufficient force has to be generated to deflect the snap (the beam) the distance required to allow the snap to engage as well as the force required to overcome the friction between the snap and the catch. The stiffness of the snap arm, the size of the angle and the coefficient of friction between the snap and the catch will determine the assembly force required. This is illustrated in Figure 9.10. In the assembly process the catch portion of the snap fit is pushed into the snap arm to engage the snap fit with a force, $F_A$ (the assembly force), by acting on the tapered surface on the engagement side of the snap. This creates a perpendicular force, $F_d$ (the deflection force), which will cause the snap arm to move downward, allowing the catch to pass over it. As the catch is pushed over the snap the force must be sufficient to deflect the snap arm and also overcome the friction between the two sliding surfaces. The assembly force required is described by Equation 9.10, where $\mu$ is the coefficient of friction between the snap material and the catch material and angle A is the angle of the engagement side of the snap, as shown in Figure 9.10.

$$\text{Assembly Force} = F_A = F_d \frac{\mu + \tan A}{1 - \mu \tan A} \tag{9.10}$$

It follows that in order to disassemble (disengage) the snap the process will be reversed, deflecting the snap from the disengagement side as defined by Equation 9.11.

$$\text{Disassembly Force} = F_\mu = F_d \frac{\mu + \tan B}{1 - \mu \tan B} \tag{9.11}$$

Again, $\mu$ is the coefficient of friction between the snap and catch material and angle $B$ is the angle of the disengagement side of the snap.

To solve these equations it will be necessary to be able to calculate the deflection force, $F_d$, that will be required. This can be accomplished using Equation 9.12, deflection force for a rectangular beam, shown below, where $h_o$ again is the thickness of the snap at its base, $b$ and $L$ are the width and length of the snap respectively, $\varepsilon$ is the strain at full deflection of the snap and $E_s$ is the strain dependent modulus of elasticity or secant modulus of the material the snap is made from.

$$\text{Deflection Force} = F_{d=} \frac{bh_0^{\,2}}{6} \; X \; \frac{E_s\,\varepsilon}{L} \tag{9.12}$$

The secant modulus is used because it gives a more realistic value of the stress-strain relationship at the strain level of interest. Figure 9.11 shows a stress-strain curve similar to the ones we looked at in Chapters 2 and 6.

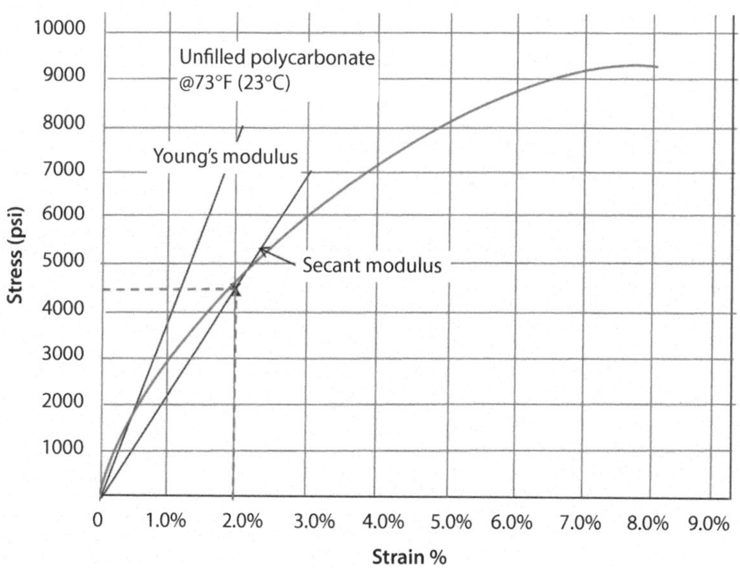


**Figure 9.11** Secant modulus.

At the start of this curve is a straight line portion where the stress is proportional to strain. The slope of the tangent to this portion of the curve is known as Young's modulus and is the modulus figure normally given in data sheets. For most plastics, this region is extremely small and to get a more accurate modulus measurement at higher strain values, we construct a secant line from the origin to the point of interest (in this case at 2% elongation). The slope of this line is known as the secant modulus. If we used the modulus of elasticity from the Young's modulus, we would have had a resulting stress much higher at the 2% strain level than the true stress as determined by the secant modulus. To calculate the secant modulus, we divide the stress by the strain at the point of interest (4600/.02) and find the modulus to be 230,000 psi. At lower strain levels, the secant modulus will approach the Young's modulus or the modulus of elasticity for the material.

We will now look at a few examples of how these equations coupled with some of the previous ones can be used in the design process. First, we are designing a polycarbonate case that will snap into a catch in its cover and will be opened and closed many times. The snap designed needs to be 7/8″ long to engage the cover and is 3/8″" wide. A rectangular cross section tapered to ½ the base thickness has been chosen for the snap profile (a tapered cantilever beam). The catch depth is 3/32″ and the engagement angle is 30°. We need to calculate the thickness of the snap at the base, $h_o$, find the force required to deflect the snap, $F_d$, and find the assembly force, $F_A$, that will be required. We know the following:

Length: $L$ = 7/8″ = 0.875″

Width of base: $b$ = 3/8″ = 0.375″

Deflection: $D$ = 3/32″ = 0.938

Ratio of $h/h_o$ = 50%

We first need to find what the base thickness must be. To do that we will use Equation 9.4

$$\varepsilon = 1.5\,\frac{\Delta h_0}{L^2 K} \tag{9.4}$$

Rearranging the equation to solve for $h_0$ we get

$$h_0 = \frac{\varepsilon L^2 K}{1.5\Delta} \tag{9.13}$$

From Table 9.2, we find the allowable strain for polycarbonate of 4%. Since our application requires multiple uses we will design for 50% of that value or 2% which = $\varepsilon$. From Table 9.1 we find the proportionality constant K for a tapered beam ratio $h/h_o$ of 50% to be 1.636. We now have all the information we need to solve for the thickness of the snap arm at the base, $h_o$. Substituting in the above equation we find:

$$h_0 = \frac{(.02)(0.75)^2(1.636)}{(1.5)(.0938)} = 0.178'' \tag{9.14}$$

Next we need to find the force that will be required to deflect the snap arm the required 0.0938″ to provide the movement needed for the arm to engage the catch. For this we will use Equation 9.9, the equation for deflection force. We have all the data we need with the exception of the secant modulus, $E_s$. We will get this from the graph in Figure 9.10 where we have already calculated the secant modulus for polycarbonate at 2% strain to be 230,000 psi. Substituting these values in the above equation-yields the deflection force:

$$\text{Deflection Force} = F_{d=} \frac{bh_0{}^2}{6} X \frac{E_s\,\varepsilon}{L} \tag{9.12}$$

$$F_d = \frac{(0.375)(0.178)^2}{6} \times \frac{(230,000)(0.02)}{0.875} \qquad F_d = 10.4 \text{ lbs.} \tag{9.15}$$

Lastly, we need to determine the amount of assembly force that will be required to engage the snap fit. To do this we will use Equation 9.10.

$$\text{Assembly Force} = F_A = F_d \frac{\mu + \tan A}{1 - \mu \tan A} \tag{9.10}$$

From Table 9.2 we find the coefficient of friction for polycarbonate on polycarbonate of 0.4 so we have all the data we need to solve the equation:

$$F_A = 10.4 \frac{0.4 + 0.577}{1 - 0.4 \, x \, 0.577} \tag{9.16}$$

$$F_A = 13.2 \text{ lbs}$$

We find that the assembly force that will be required is 13.2 pounds.

In our next example, we will design a simple rectangular cross-section snap fit to hold a housing cover on the side of a business machine. We have chosen our material to be PBT. From the material supplier, we were given the following data, permissible strain $\varepsilon$ = 4.4%, coefficient of friction for this grade of PBT on PBT $\mu$ = 0.30 and the secant modulus at 4.4% strain, $E_s$ = 4800 Mpa. We have 25 mm from the base where the snap will attach to the machine to the catch of the snap. The width is 6 mm and the thickness has been calculated with Equation 9.2 to be 2.2 mm. The engagement angle is 30°. The housing is located in an area with a lot of activity. To prevent accidental disengagement of the housing, what is the disengagement angle B that will cause the disassembly force to be at least double the assembly force?

First, again using Equation 9.12 the deflection force required is calculated as follows:

$$F_d = \frac{(6.0)(2.2)^2}{6} \times \frac{(4800)(0.044)}{25.0} \tag{9.17}$$

$$F_d = 30.7 \text{ newtons}$$

Now that we know the deflection force we can calculate the assembly force required from Equation 9.10.

$$\text{Assembly Force} = F_A = F_d \frac{\mu + \tan A}{1 - \mu \tan A} \tag{9.10}$$

$$F_a = 30.7 \frac{0.3 + 0.577}{1 - 0.3 \, x \, 0.577} \tag{9.18}$$

$$F_A = 32.5 \text{ newtons}$$

The requirement for the disassembly force, $F_U$, is that it is at least double the assembly force, so we know that $F_u$ has to be at least 2 × 32.5 N or

65 N. Looking at Equation 9.11 for the disassembly force, the forces $F_D$ and $F_u$, and the coefficient of friction $\mu$ are known and we need to solve for the tangent of the disengagement angle $B$, so we can find the angle required.

$$\text{Disassembly Force} = F_u = F_d \frac{\mu + \tan B}{1 - \mu \tan B} \quad (9.11)$$

We do this by rearranging the above equation and solve for Tan B:

$$Tan\ B = \frac{(F_u - F_d\mu)}{F_u\mu + F_d}$$

$$Tan\ B = \frac{(65.0 - 30.7 \times 0.3)}{(65.0 \times 0.3 + 30.7)} = 1.1122 \quad (9.19)$$

$$\text{Angle } B = 48°$$

If we make the disengagement angle 48° the force required for disassembly will be 65 newtons.

When angle B approaches 90°, the catch can no longer slide over the detent and the disassembly force will be determined by the area of the tip of the snap and the shear stress of the plastic being used.

If a snap fit is meant to be used numerous times, a couple of feature enhancements should be considered as part of the design. When trying to disengage a snap, the hook portion must be pushed back. If the disassembly pressure is high, often a metal tool, such as a screw driver tip, will be used, which can damage the tip of the hook. A solution to this is shown in the left-hand side of Figure 9.12, where an extension has been added to the top of the snap which serves as a release lever for the

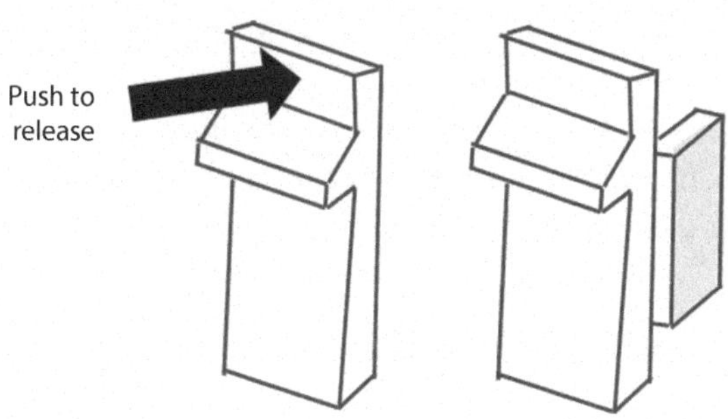


**Figure 9.12** Snap fit with release lever for disassembly.

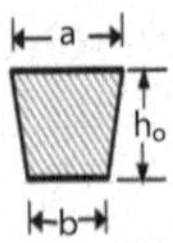


Cross section constant over beam length L

$$\text{Deflection} = \Delta = \frac{a+b}{2a+b} \times \frac{\varepsilon L^2}{h_o} \qquad \text{Deflection force} = F_d = \frac{h_o^2}{12} \times \frac{a^2+4ab+b^2}{2a+b} \times \frac{E_s \varepsilon}{L}$$

Tapered cross section over beam length L from $h_o$ at base to h/2 at the tip

$$\text{Deflection} = \Delta = 1.64x \frac{a+b}{2a+b} \times \frac{\varepsilon L^2}{h_o} \qquad \text{Deflection force} = F_d = \frac{h_o^2}{12} \times \frac{a^2+4ab+b^2}{2a+b} \times \frac{E_s \varepsilon}{L}$$

**Figure 9.13** Trapezoidal non-rectangular cantilevered beam equations.

undercut and also provides some mechanical advantage to ease disassembly. When a snap is used multiple times, there is always the risk of overstressing the snap either by pushing it too far back and greatly exceeding its strain limit or by slightly exceeding the strain limit multiple times. This can be prevented by designing a backstop as a permanent part of the design, as illustrated on the right-hand side of Figure 9.12. In a clever design the backstop can serve as an additional guiding or mating feature or another functional part.

## 9.4 Non-Rectangular Cantilevered Beams

For trapezoidal non-rectangular cantilevered beams the equations in Figure 9.13 [35] can be used in place of Equations 9.3 and 9.12.

For these and other more complex geometries, the most accurate results can be developed through finite element analysis (FEA) calculations. The calculations based on classical beam theory use simplifying assumptions such as the end of the beam is assumed to be attached to a totally rigid wall that will not deflect as the beam is deflected. Actually, movement of the beam will cause deflection in the body it is attached to. This deflection will reduce the strain on the beam, so strains calculated by the classical methods will usually be higher than those actually encountered in typical components. FEA is able to take these variations into account.

In addition to a FEA analysis, when creating a part design a number of online snap fit calculators are available and several material suppliers will assist with designs in their materials.

## 9.5 Effects of Stress Concentration

When designing a cantilever snap fit deflection stresses tend to concentrate at the base of the snap arm where it joins the part wall or support

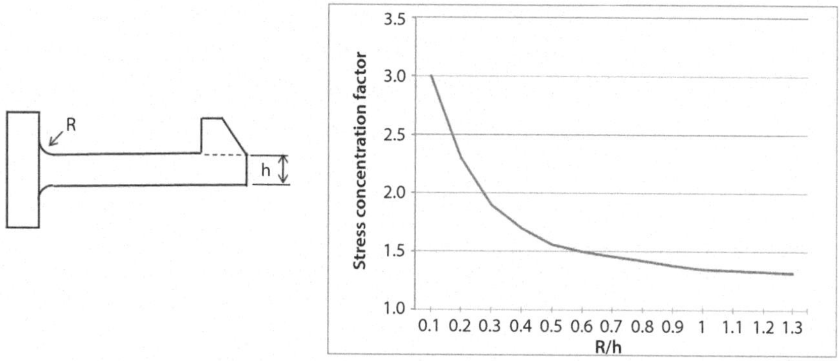


**Figure 9.14** Effect of base radius on stress concentration.

structure. Sharp corners in this area can produce cracks from repeated use of the snap fit, particularly with rigid plastics that have low elongations that are typical of glass-reinforced materials. This condition can be greatly improved by providing generous radii at the intersection point. Figure 9.14 [36] illustrates the effect that the base radius has on stress concentration, where the ratio of base radius to beam thickness, R/h, is plotted against relative stress. R/h ratios of 0.5 to 0.6 seem to provide the optimum reduction with only marginal improvements obtained with larger ratios. However, even at an R/h ratio of 0.5, you can get a thick section at this radius that can result in voids in the molded part that will severely weaken the structure. This is more of an issue as the snap arm becomes very thick. The designer needs to compromise here, providing as large a radius as possible that can be molded without voids or sinks. Minimally, the radius should be 0.015″ (0.4 mm).

## 9.6 Annular Snap Fits

Annular snap fits are commonly used in many applications, including food containers, pill containers, pens, and housings. Annular or cylindrical snap fits are generally stronger, but due to their continuous nature require greater assembly force than a cantilevered design.

Figure 9.15 illustrates a typical annular snap fit and details the key dimensions required for creating a design. The snap fit is made of two parts; the top female portion is called the hub, and the bottom male portion is called the shaft. The height of the undercut H (2 × H/2) will deform the parts upon assembly, creating a strain. Although both parts will be affected, for simplicity we will assume all the deformation occurs on the hub corresponding to the total depth of the undercut $H$, and the strain $\varepsilon$ will be described by the following equation [37]:

$$E = \frac{H}{D_G} \times 100\% \tag{9.20}$$

The assembly force and the pullout force for annular snap fits cannot be as easily calculated as a cantilever snap because the total area deformed during assembly is not known. The calculations below will give a rough approximation of them.

As the snap fit is pushed together it will create a pressure on the joint described by the following equation [37]:

$$P = \frac{H}{D_K} \times E_s \times \frac{1}{K} \tag{9.21}$$

where $P$ is the pressure, $E_S$ is the material secant modulus and $K$ is a geometry factor that depends on the dimensions of the snap fit and is described by Equation 9.22 [37].

$$K = \frac{(D_A/D_G)^2 + 1}{(D_A/D_G)^2 - 1} + 1 \tag{9.22}$$

The assembly force (or the pullout force) can be calculated using the following equation [37]:

$$F = P \times \pi \times D_G \times 2b \times \frac{\mu + \tan A}{1 - \mu \tan A} \tag{9.23}$$

The following example will demonstrate how to apply these equations. In designing a small cylindrical container similar to the one shown in Figure 9.15, the outside diameter $D_A$ is 50 mm, the wall thickness is 2 mm, typical making dimension $D_G$ 46 mm, and $H$ is 0.23 mm. Angle $A$ has been set at 30° and $b$ can be calculated to be 0.2 mm. The material selected is acetal with a coefficient of friction $\mu$ = 0.2 and $E_S$ = 1900 N/mm$^2$. We need to calculate the assembly force that will be required to see if it is reasonable and also determine if acetal is an acceptable choice for the strain that will be generated.

First we will look at the strain using Equation 9.20.

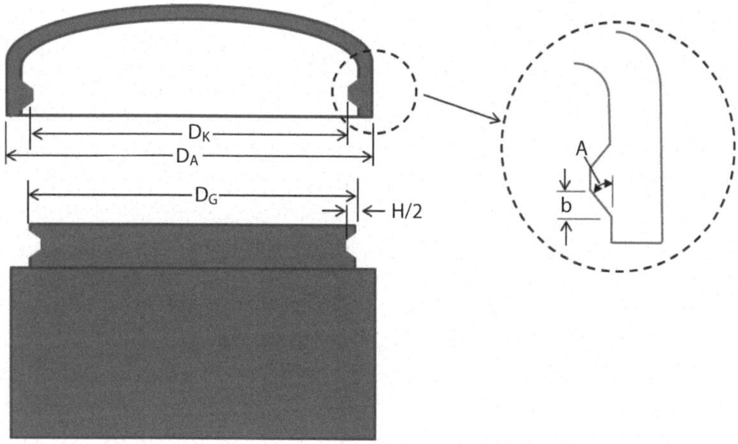


**Figure 9.15** Annular snap fit.

$$\varepsilon = \frac{H}{D_G} \times 100\% = \frac{0.23}{46} \times 100\% = 0.5\% \qquad (9.24)$$

The strain of 0.5% is well within the allowable stress for acetal from Table 9.2.

Next, we need to calculate *P*, but to do that we need to determine the geometry factor *K* from Equation 9.22.

$$K = \frac{(D_A/D_G)^2 + 1}{(D_A/D_G)^2 - 1} + 1 = \frac{(50/46)^2 + 1}{(50/46)^2 - 1} + 1 = 13.0 \qquad (9.25)$$

Using the value we determined for *K*, *P* can now be calculated.

$$P = \frac{H}{D_K} \times E_s \times \frac{1}{K} = \frac{0.23}{45.8} \times 1900 \times \frac{1}{13.0} = 0.7 \qquad (9.26)$$

Substituting the value for *P* into Equation 9.23, we can calculate the required assembly force *F*.

$$F = P \times \pi \times D_G \times 2b \times \frac{\mu + tanA}{1 - \mu tanA}$$
$$= 0.7 \times 3.14 \times 46 \times 2 \times 0.2 \times \frac{0.2 + 0.577}{1 - (0.2x0.577)}$$
$$= 37.1N \tag{9.27}$$

The assembly force required of 37.1N (8.3 lbs) appears to be reasonable. If a lower value were desired, the undercut, H, could be reduced slightly or the container wall thickness could be thinned. If the undercut has to be made too small to achieve the desired force, an alternative is to segment the undercut with alternating areas of undercut and smooth wall around the circumference.

## 9.7 Manufacturability

When developing snap fit design, there are several things that need to be considered to improve manufacturability and reduce costs. The importance of incorporating radii but also minimizing thick sections has already been discussed. When designing a snap, consideration should be given to where the hook is located with relation to the rest of the part. In the part shown in Figure 9.16, the hook is located over a solid wall, creating an undercut. The only way to produce this in a mold is by incorporating a slide or side core, as shown to the left of the part, or by adding a lifter in the tool, both of which will add to the tool cost. A better alternative, if the part allows it, is shown in Figure 9.17, where a hole through the part is provided under the hook, which will allow forming the hook in press motion with a core from below, which

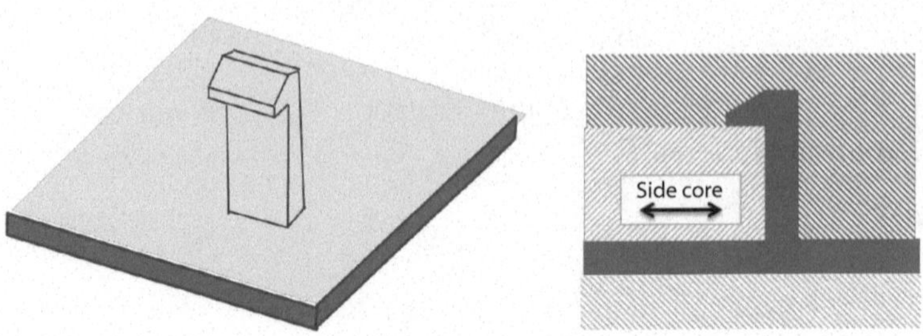


**Figure 9.16** Snap design with hook located over a solid wall.

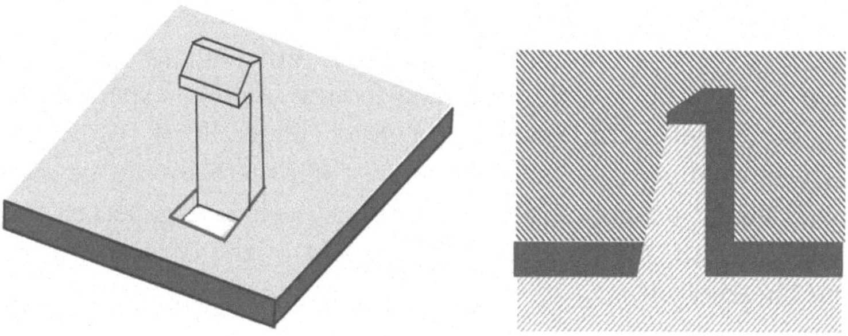

**Figure 9.17** Snap design where the hole through part is provided under the hook.

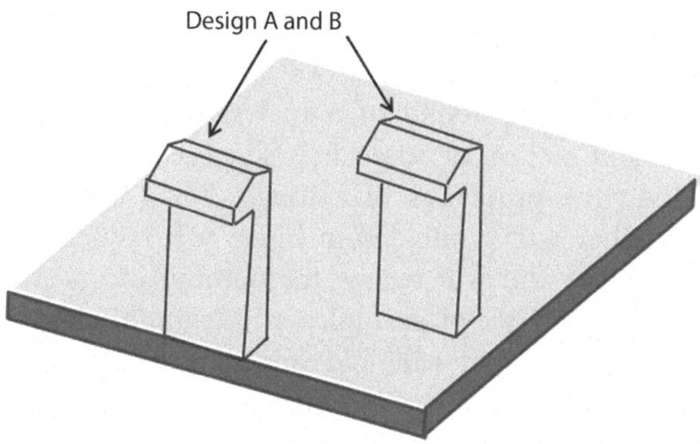


**Figure 9.18** Alternative snap designs.

is a less expensive option. When doing this it is important to be aware of where the gating of the part will be. Ideally, the gate will be somewhere behind the snap fit so that any weld line that forms will be on the hole opposite the snap. If the weld line occurs in the snap arm itself, it will significantly weaken it.

Another alternative to consider is illustrated in Figure 9.18. Design B is similar to the first design we looked at where the hook is located over the wall of the part. If the snap can be moved to an outside wall so the hook portion is no longer trapped behind the wall, as shown in design A, the part can now be molded all in press motion without requiring any side cores of lifters.

Lastly, designing the arms as long as reasonable will reduce the amount of stress on the arm and prolong its life.

When making a cylindrical snap-fit, the molded raised lip may be removed from the mold by snapping (pushing) off the core with the ejector system. This requires a material with good elastic recovery characteristics that will allow the part to be momentarily stretched without permanent deformation. Alternatively, the undercut can be made with an internal collapsible core. The tooling is more expensive for this option but it will permit molding a much larger lip if required. If the lip is to be snapped (pushed) off of the core, there should be an ample radius designed on both sides of the lip, as well as a generous taper on the core above it, to ease the lip sliding out of the core and to prevent it from shearing off.

## 9.8 Plastic Hinges

Many thermoplastics can be used to incorporate an integral hinge as part of the design. This integral plastics hinge is sometimes referred to as a "living hinge." The ability of a material to orient itself when it flows across a thin section or when it is squeezed into a thin section is key to achieving the properties of a hinge. Being able to join two or more parts together with a molded-in hinge will reduce the part count that has to be made and will reduce the amount of assembly required. Figure 9.19 shows a couple of examples of integral hinges. On the left is an electrical connector that will fold or hinge two halves together to capture the wires, which will be held together by a snap fit. On the right is a pill box where each daily compartment has its own hinged lid. It also has a snap fit on the front of each individual lid.

Hinges can be produced by molding or extrusion and also by cold forming or coining of the material to be hinged.

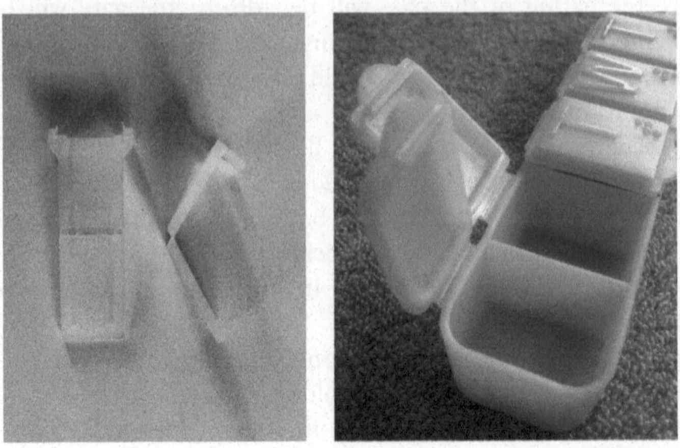

**Figure 9.19** Examples of molded-in hinges.

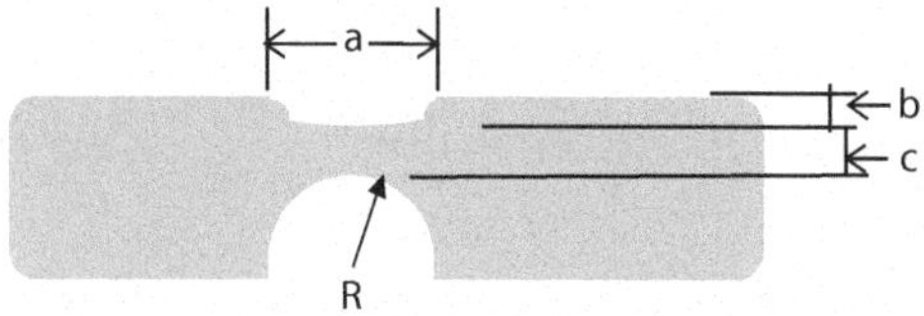


| a | .040–0.80″ | 1–2 mm |
|---|---|---|
| b | .006–.012″ | 0.15–0.3 mm |
| c | .010–.015″ | 0.25–0.4 mm |
| R | .020–.040″ | 0.5–1.0 mm |

**Figure 9.20** Typical polypropylene hinge designs.

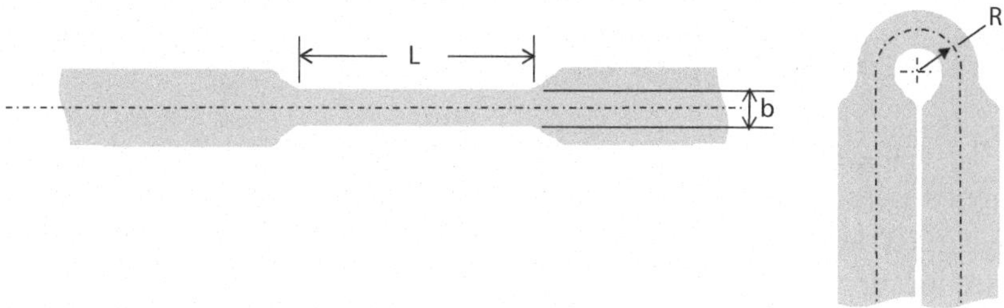


**Figure 9.21** Plastic hinge in bending.

Injection molding is the most common way of producing a hinge. Polypropylene is the material most commonly used for hinges but polyethylene, nylon, or most materials with good elongation will work. Filled materials are a problem as the fillers will interfere with the orientation of the plastic in the hinge area and can cause weak spots that will fail in the thin sections. Even some color concentrates and pigments can cause issues if not uniformly dispersed. It is good practice to consult with your material supplier for recommendations.

Figure 9.20 shows some design ranges for an integral polypropylene hinge.

As the hinge is bent it will develop stresses on the radius of the bend. Figure 9.21 illustrates a hinge being bent.

Equations 9.28, 9.29 and 9.30 [38] calculate the bending stresses and strains in a bent hinge:

$$R = \frac{L}{\pi} \text{ For } 180° \text{ bending} \tag{9.28}$$

$$\varepsilon = \frac{\pi b}{2L} \quad \text{strain in extreme fiber} \tag{9.29}$$

$$\sigma = \frac{\pi b E}{2L(1 - \upsilon^2)} \text{ stress in extreme fiber} \tag{9.30}$$

where $\varepsilon$ = strain, $s$ = stress, $E$ = the modulus of elasticity and $\upsilon$ = Poisson's ratio. With these calculations we can determine acceptable ratios of hinge thickness ($b$) to hinge length ($L$) using Equation 9.31 [38].

$$\frac{b}{L} < \frac{2\sigma\varepsilon}{\pi E}\left(1 - \upsilon^2\right) \text{ for elastic behavior} \tag{9.31}$$

To achieve the best hinge properties, you want the molecules to be aligned perpendicular to the hinge axis. This is generally achieved with high melt temperatures and a fast fill time through the hinge section. Gate location and sizing is very important because it has a major effect on molecular orientation across the hinge. Care must be taken to prevent weld lines or gas entrapment in the hinge area. Fan gates can be very effective for this.

When extruding it is difficult to get the proper orientation in the hinge area in the direction of extrusion flow. The best results are achieved with secondary forming at right angles to the extrusion flow.

Hinges can also be formed after the part is molded or extruded. When parts get larger it is sometimes impractical to force material from one half of the part across a thin hinge and fill the other half. This can be done with a small hydraulic or toggle press with bars the shape of the hinge which will squeeze the part in the hinge to the desired thickness. The bars are usually heated (about 275 °F for polypropylene). This process will work faster if the part is formed immediately after molding while it still retains some heat. Without heat this process is called "coining" and works well with some engineering materials such as nylon or acetal.

# 10

# Plastic Gears

Plastic gears first made their appearance over 60 years ago when nylon and acetal thermoplastic materials were applied to low power gear applications, such as lawn sprinklers and inexpensive clocks, taking advantage of the moldability and natural lubricity of these materials. Gears machined from phenolics and other thermosets allowed use at higher operating temperatures and were generally resistant to lubricants used.

Since then improvements in engineering polymers and manufacturing processes have allowed plastic gears to establish themselves as serious competitors and often the material of choice over traditional metal gears in a wide variety of applications. They have moved from low power, low precision applications into precise motion and higher power transmission applications.

The design flexibility inherent in today's molded plastics allows for producing gears with novel tooth shapes that would be difficult to form at a reasonable price in metals. Plastic gears typically permit a 50% to 90% cost reduction versus stamped or machined metal gears because they usually require no secondary finishing [39]. They can also incorporate other features that might be needed to aid assembly, which can further reduce costs.

The materials used often require no lubrication because of the low coefficients of friction of the materials themselves, but also can be compounded with internal lubricants, such as PTFE or silicone, to provide self-lubrication. This makes them an excellent choice for applications where lubricants must be avoided such as copy machines, computer printers and child's toys. Additionally, materials are available that are resistant to many common lubricants and corrosive operational environments. Plastic gear teeth will deflect under load, which will better distribute localized loads caused by misalignment and tooth geometry errors, and will also help absorb impact loads.

The inherent low weight (1/3 to 1/8 that of a comparable metal gear) and inertia of plastic materials reduces the power requirements and the low modulus results in lower noise generation.

Similar to previous discussions, the unique viscoelastic behavior of plastic materials must be accounted for when designing a gear. Load carrying and wear data is not as extensive as that which exists for metals but the experience base with plastics gears has grown and material supplier's data and CAD design software is now available that allows reasonable development of plastic gear designs. Plastic gearing is being used in higher precision and more demanding service applications as new materials are developed for gear applications and designers continue to push the limits to take advantage of the cost/performance benefits that plastics offer.

## 10.1 How Gears Work

Gears are used to transmit motion and work either with another gear or with a shaft. Gears must be used at least in pairs in order to do work and frequently a number of gears will be used in a gear train. When the teeth of two gears are in contact the gears are said to be "meshed" and the rotation of one of the gears will drive the second gear and cause it to rotate also. The gear through which the rotational force is initiated is called the "drive gear" and the gear to which it transmits force is called the "driven gear." When the two gears have different diameters, the larger gear will be referred to as "the gear" and the smaller one will be called "the pinion." In operation the pinion will rotate faster than the gear.

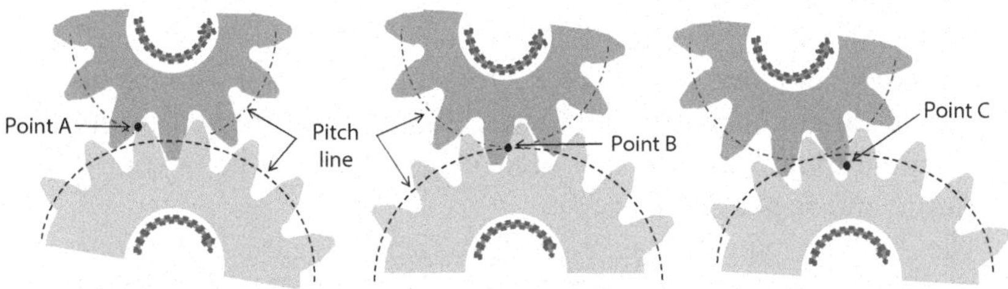


**Figure 10.1** Gear meshing process.

When the gear teeth come in contact with each other, as each tooth is loaded it acts like a cantilever beam supported on one end. The contact force will try to bend and shear the beam or tooth. As the two gear teeth mesh and then unmesh, frictional forces which develop as they slide over one another will create additional shear and compression forces. Figure 10.1 illustrates the gear meshing process. In the first figure at the left, the driven gear tooth on the smaller pinion gear makes its first contact with a tooth on the larger gear at point A. This contact takes place below the pitch point (the point of tangency between two pitch circles of gears in mesh – more on his concept later) of the pinion and below the pitch point of the gear. Note that the pair of gear teeth in front of these teeth is also in contact, reducing the loading on each tooth pair. In the middle figure the gear has rolled forward and the face of the two teeth in contact have rolled and slid across each other until the two teeth at point B are in contact at the pitch line of both gears. This combination of rolling and sliding action subject the gear tooth face to a combination of compressive and tensile forces called contact forces that cause tooth wear. At this point, B, the single tooth pair is carrying the entire load and this is the point of maximum tooth bending stress. On the right-hand side of the figure the gears rotate further and the tip of the drive gear is now in contact with the driven gear below its pitch point as the gears move to unmeshing with one another (point C). At this point the tooth loading is again shared by at least two gear pairs. The length of arc of the pitch circle from point A to point C is called the arc of action.

When selecting a gear material the designer has to consider flexural strength and stiffness as well as coefficient of friction, as the gear tooth needs to provide resistance not only to the bending stresses but also the wear caused by the contact stresses.

## 10.2 Types of Gears

There are a number of different types of gears that can be used to address a variety of motion problems. Plastic gears are used in many of these applications and are also used in gear trains with a combination of plastic and metal gears.

The simplest types of gears are spur gears (Figure 10.2). Spur gears operate on axes that are parallel to one another. The teeth of a spur gear are pointed perpendicularly from the axis of rotation and the face or "flank" of the gear tooth is parallel to the rotational axis. If the gear teeth point away from the axis, it is an external spur gear, if they point toward the axis, it is an internal spur gear. Spur gears are the simplest gears to design, manufacture and assemble. When spur gears are operated, they will develop only radial loads.

Face gears are a special adaptation of a spur gear (Figure 10.3). In a face gear, one gear has the teeth molded or cut into the face of the gear. The teeth in the face gear will point parallel to the gear axis. The face gear will mate with either a spur gear or a helical gear. Face gears permit the transfer of power at an angle, usually with the two gear axes at 90° to each other.

Helical gears are similar to spur gears but the tooth flank (face) is at an angle to the axis of rotation of the gear instead of being parallel to it. The angle the tooth is at in relation to the axis is called the helix angle. The helix angle will cause the gear to develop both a radial load and a thrust load. Helical gears will be able to handle a larger load than an equivalent spur gear and will run more quietly, but are more difficult to manufacture and assemble.

In Figure 10.4 we see an example of simple helical gears and a double helical gear. A double helical gear is made by molding or mounting two

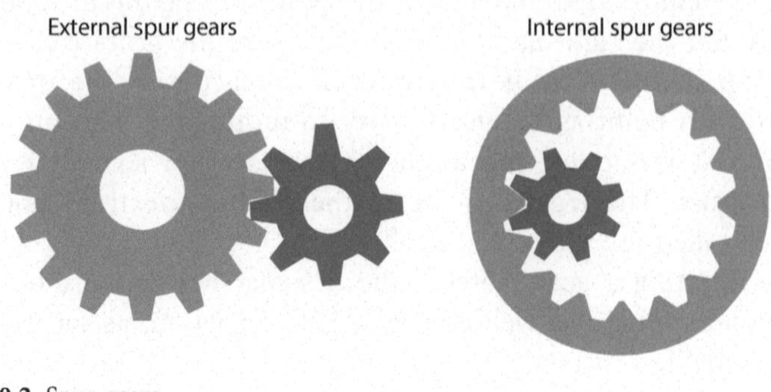


**Figure 10.2** Spur gears.

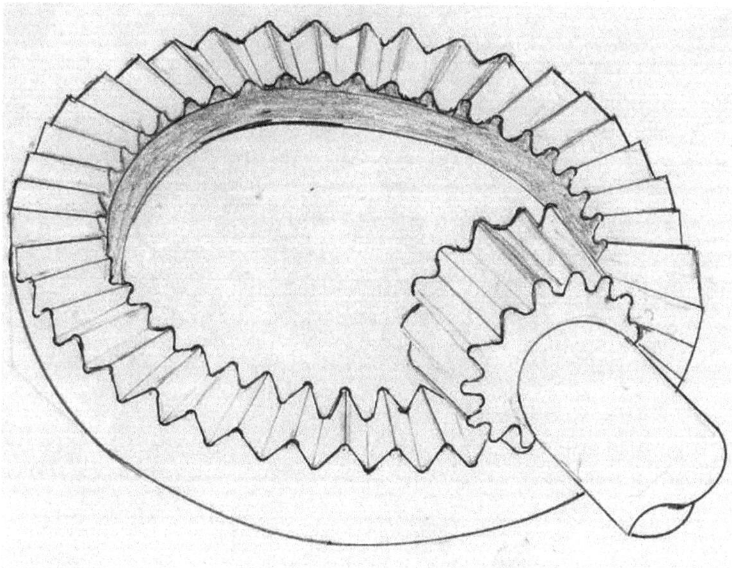

**Figure 10.3** Face gear.

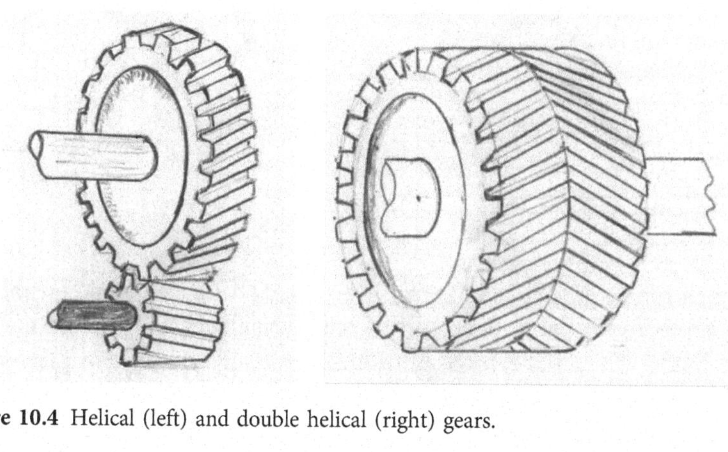

**Figure 10.4** Helical (left) and double helical (right) gears.

helical gears that have opposite helix angles, for example, +15° and −15°. A double helical gear will cancel out the thrust load. Double helical gears are also sometimes called herringbone gears.

Bevel gears (Figure 10.5) are used to transmit motion at an angle, usually at 90°, although that can work at a range of different angles. The gears are conical with the teeth being tapered both in height and thickness. The tooth dimension specified is based on the large end of

**Figure 10.5** Bevel gear.

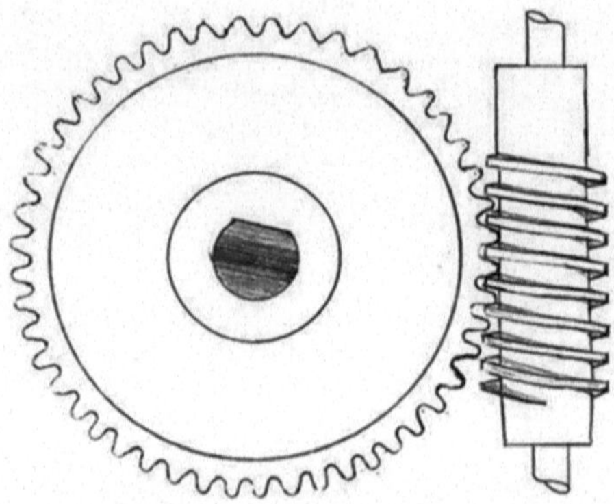

**Figure 10.6** Worm gear.

the tooth but strength calculations are typically done on the central section of the tooth [40]. Bevel gears will be subject to both radial and thrust loads. They have to be mounted precisely. Their complex shape makes them a good candidate to be injection molded.

Worm gears are another gear arrangement which transmits motion at an angle, again usually 90°. Worm gears (Figure 10.6) are made up of a gear with a screw thread (gear on the left) called the worm and the gear it mates with called the worm gear. The simplest worm gear is a simple helical gear coupled with a worm. This is called a non-enveloping worm gear. This gear set design produces line contact between the gears and places a high level of stress on the gear teeth. The point contact makes this design not particularly good at carrying high loads but it is relatively easy to manufacture and assemble.

**Figure 10.7** Compound gear.

The load carrying capacity can be improved with a single-enveloping design. In this design, the worm gear will have a concave tooth design that wraps around the worm thread. This increases the area of tooth contact and will increase the load carrying capability by a factor of 2–3 times [41] over a simple helical gear. It is more complex to make.

In a double-enveloping gear design the worm itself will also have a curved profile to better match the curvature of the worm gear. This further increases the load carrying capability of the gear set but is even more difficult to manufacture and assemble.

When designing a worm gear set, it is not unusual to couple a metal worm with a plastic gear, particularly if the gear set is to run at a high speed. The frictional heat generated by a worm gear set can be quite high and the metal worm will help to dissipate this heat.

Compound gears or cluster gears (Figure 10.7) are when two or more gears are made as one piece. The gears that make up a compound gear will usually differ in size and have a different number of teeth. The purpose of this arrangement is to speed up or slow down the final output from the input speed. Obviously, plastic molding is a great way to manufacture these types of gears.

## 10.3 Terminology

Before we proceed any further, we should review some of the terminology that is used in defining gears and gear operation. Many of the following definitions will be further illustrated in Figures 10.8 and 10.9.

> Addendum – the depth of tooth space above the pitch line.

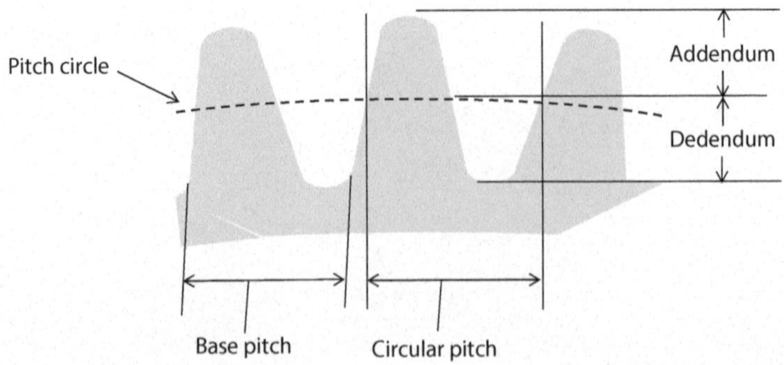


**Figure 10.8** Gear terminology 1.

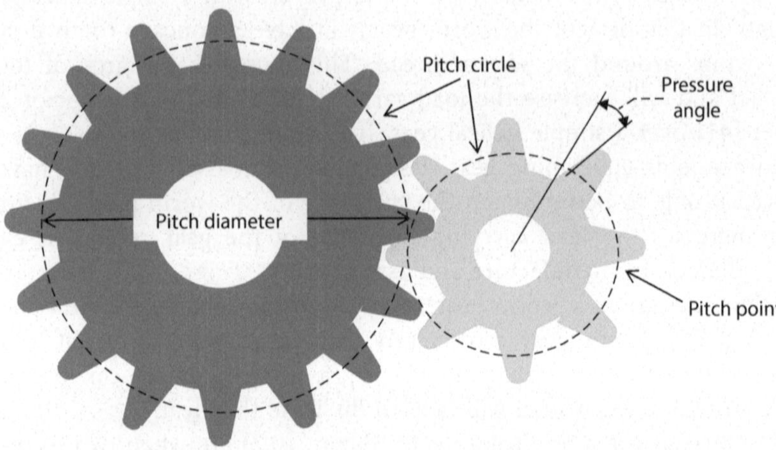


**Figure 10.9** Gear terminology 2.

Arc of Action – arc of the pitch circle through which a tooth travels from the first point of contact with the mating tooth to the point where contact ends.

Backlash – the tangential clearance between two meshing teeth. Some backlash is necessary for smooth rotation of meshed gears.

Base Circle – the circle from which an involute tooth curve is developed.

Base Pitch – the pitch on the base circle for an involute gear.

Circular Pitch (p) – the length of arc on the pitch circle between corresponding points of two adjacent gear teeth.

Diametrical Pitch (P) – the ratio of number of teeth to the number of inches of pitch diameter. It equals the number of teeth per inch of pitch diameter.
Dedendum – the depth of tooth space below the pitch line.
Gear Ratio – the number of teeth in the gear divided by the number of teeth in the pinion.
Module (M) – the pitch diameter divided by the number of gear teeth.
Pitch Circle – a circle the radius of which is equal to the distance from the gear axis to the pitch point.
Pitch Diameter ($P_d$) – the diameter of the pitch circle. It is equal to the total number of teeth in a gear divided by the diametrical pitch.
Pitch Point – the point of tangency between two pitch circles of gears in mesh.
Pressure Angle – the angle between the tooth face and a radial line at its pitch point.

## 10.4 Gear Tooth Loading

The gear teeth are what allows the gear to provide motion and do work. Gear teeth transmit the torque load generated at one gear shaft and transmit it to another. The ability of a gear to carry a given load is largely determined by the strength of the individual teeth.

The classical method of calculating the bending stresses in a spur gear tooth is the Lewis equation. This method assumes the full load is carried by the tooth loaded at the pitch line, acting like a cantilever beam. It is derived from the bending stress equation (Equation 6.1 reprinted below) from Chapter 6:

$$S = \frac{Mc}{I} \tag{6.1}$$

Modifying this equation to fit the geometry of a gear tooth results in the gear tooth stress equation:

$$S_b = \frac{FP_d}{f\gamma} \tag{10.1}$$

where $S_b$ = the bending stress of a tooth loaded at the pitch line, $F$ = the tangential loading at the pitch line, $P_d$ = the pitch diameter of the gear, $f$ = the face width of the gear tooth, and $\gamma$ = the Lewis form factor for gears loaded at the pitch point. The most severe tooth loading occurs when the number of pairs of teeth in mesh approach 1 and the teeth are tangentially loaded at the pitch line [42].

Table 10.1 gives some sample Lewis form factors [43] for gears with a 14–1/2-degree pressure angle full-depth tooth, a 20-degree pressure angle full-depth tooth, and a 20-degree pressure angle with a stub-form tooth. A 20-degree pressure angle is commonly used in plastic gear design. As can be seen from the table, since the form factor is in the denominator of Equation 10.1, the 20° angle will give a lower stress level than the 14–1/2° angle. The form of the tooth will obviously affect the Lewis factor. If we went to a 20° stub design (a shorter, stubbier tooth than the full depth design) the corresponding Lewis factors would be greater, resulting in a lower calculated stress.

When selecting a tooth form a number of factors must be considered. A larger pressure angle will give a thicker and stronger tooth, but a thinner tooth can be molded to a more accurate profile, which will result in a more quiet operation. As the pitch velocity increases, this becomes increasingly important. A stub tooth form, which is different from the standard tooth form in tooth depth, will carry more load, but the full depth tooth with more surface area to distribute the shear and compression forces will generally provide a longer gear life.

Let's look at an example of a tooth bending stress calculation. Assume we want to design a 36 tooth gear with a 20° pressure angle and full depth design. The gear will have 8 teeth per inch (the diametrical pitch) and the face width will be 0.5″. The gear will be driven by a 1/2 horsepower motor at 75 revolutions per minute. Acetal has been suggested as the material to use and we want to be sure it has adequate strength.

We will use Equation 10.1 above. We are given the tooth face width, f, of ½″. From the Lewis table we can find the appropriate form factor $\gamma$ of 0.64 (interpolation between 34 and 38 teeth). We need to calculate the pitch diameter ($P_d$) and the tangential tooth loading at the pitch line, F.

**Table 10.1** Lewis tooth factory $\gamma$ (load near pitch point).

| Number of Teeth | 14–1/2° | 20° Full Depth | 20° Stub |
|---|---|---|---|
| 16 | | | 0.578 |
| 17 | | 0.512 | 0.587 |
| 18 | | 0.521 | 0.603 |
| 19 | | 0.534 | 0.616 |
| 20 | | 0.554 | 0.628 |
| 22 | | 0.559 | 0.648 |
| 24 | 0.509 | 0.572 | 0.664 |
| 26 | 0.522 | 0.588 | 0.678 |
| 28 | 0.535 | 0.597 | 0.688 |
| 30 | 0.540 | 0.606 | 0.698 |
| 34 | 0.553 | 0.628 | 0.714 |
| 38 | 0.566 | 0.651 | 0.729 |
| 43 | 0.575 | 0.672 | 0.739 |
| 50 | 0.588 | 0.694 | 0.758 |
| 60 | 0.604 | 0.713 | 0.774 |
| 75 | 0.613 | 0.735 | 0.792 |
| 100 | 0.622 | 0.757 | 0.808 |
| 150 | 0.635 | 0.779 | 0.830 |
| 300 | 0.650 | 0.801 | 0.855 |
| rack | 0.660 | 0.823 | 0.881 |

Knowing the diametrical pitch and the total number of teeth in the gear we can calculate $P_d$ as follows:

$$P_d = \frac{\text{Total teeth on gear}}{D_p} \tag{10.2}$$

Substituting we get:

$$P_d = \frac{36}{8} = 4.5'' \tag{10.3}$$

Next we need to calculate the tangential tooth loading, F. We start by finding the torque generated by the motor driving the gear.

$$T = \frac{P}{\omega} \tag{10.4}$$

where $T$ equals torque, $P$ equals power in inch pounds and $\omega$ is equal to the angular velocity at the pitch diameter in radians per second.

To convert the horsepower from the motor into inch pounds we use Equation 10.5.

$$P = hp \times 550\ ft - lbs/hp \times 12''/ft \tag{10.5}$$

Substituting we get:

$$P = {}^{1}/_{2} \times 550\ ft - lbs/hp \times 12''/ft = 3300\ inch - \text{lbs} \tag{10.6}$$

To find the angular velocity ω, we can convert the gear rpm (N) as follows:

$$\omega = N \times \frac{2\Pi}{60} \tag{10.7}$$

$$\omega = 75 \times 2(3.14)/60 = 7.85\ \text{radians/sec}$$

Now substituting in Equation 10.3 we can calculate the torque as follows:

$$T = P/\omega = 3300\ \text{inch lbs}/7.85\ \text{radians/sec} = 420.1\ \text{inch lbs} \tag{10.8}$$

Next, we have all the information we need to calculate the tangential tooth loading, $F$, using equation 10.9 below:

$$F = \frac{2T}{P_d} \tag{10.9}$$

$$F = (2) \times (420.1)/4.5 = 186.7 \text{ lbs}$$

Finally, we have all the information to solve Equation 10.1

$$S_b = \frac{FP_d}{fY} \tag{10.1}$$

$$S_b = (186.7)(4.5)/(0.5)(0.64) = 2626.1 \text{ psi}$$

We now know the stress level in the gear tooth and we need to determine an appropriate safety factor that will take into consideration factors such as operating environment, lifetime cycles, velocity, tolerances, and pitch line velocity. This is not straightforward with the Lewis method, but a good approximation to start with is 1/3 of the material's ultimate tensile strength. If we look at the tensile strength of an unfilled acetal copolymer (Celcon® M90 ™) we find it listed at 8800 psi. A third of this would be 2933 psi, so it seems as if this material would be adequate from a strength standpoint. If a little more safety factor was desired you could consider increasing the tooth width. The only certain way of determining suitability for the application is by prototyping and testing.

As mentioned above, pitch line velocity can add to tooth stress by causing impact at initial tooth contact. The Barth Velocity Factor ($K_v$) is a method to account for the effects of pitch velocity on tooth stress [44]. This factor is calculated by Equation 10.10:

$$K_v = \frac{1200 + v}{1200} \tag{10.10}$$

where $v$ is the velocity at the pitch diameter in feet per minute. For our example above we can calculate $K_v$ as follows:

$$K_v = \frac{1200 + (RPM \times \pi P_d)/12}{1200} = \frac{1200 + (75 \times \pi \times 4.5)/12}{1200} = 1.074 \tag{10.11}$$

This would increase our calculated $S_b$ above by 7%, but will leave one less factor to be included in the safety factor.

There are a number of refinements to the Lewis equation and one that I believe is worth mentioning here is the AGMA bending stress formula (American Gear Manufacturers Association www.agma.org).

$$\sigma_b = \frac{FP_d}{fY_j} \frac{K_a K_s K_m K_B}{K_v} \tag{10.12}$$

where $\sigma_b$ is the bending stress, $Y_j$ is the Lewis form factor corrected for several geometry factors, $K_a$ is an application factor that accounts for shock, $K_s$ is a size factor, $K_m$ is a load distribution factor, $K_B$ is a rim thickness factor, and $K_v$ is a dynamic factor which is a modified Barth velocity factor [45].

For gear types other than spur gears the bending stress formula needs to be modified. For example, to calculate the bending stress of a helical gear a good approximation is given in Equation 10.13.

$$F = m\frac{R_a - b}{R_a} \quad y b \sigma_b \tag{10.13}$$

$$Z_v = \frac{Z}{cos\delta} \tag{10.14}$$

where m = module(mm), $R_a$= outer cone diameter (mm), b = tooth width (mm), $\gamma$ = form factor at pitch point, obtained from Table 10.1 by calculating # of teeth of equivalent spur gear from equation 10.14, $\sigma_b$ = allowable bending stress, $Z_V$= number of teeth equivalent spur gear , and $\delta$ = pitch cone angle(degrees).

## 10.5 Contact Stress

In the section above on how gears work we introduced the concept of contact stresses. As the gears rotate causing the teeth to mesh and unmesh, this action generates stresses due to the contact of the faces of the gear teeth and their relative motion to one another. These stresses can lead to wear or failure on the surface of the gear teeth. The gears must be designed to keep these stresses within the stress limit of the material.

As the gear teeth contact one another, the contact area is initially a point or line. Under the action of the load, the teeth will deform and a contact area will be formed. In 1880 Heinrich Hertz developed a model for predicting contact area, pressure and the induced stress between two lenses. This model is known as the Hertz theory of contact stress and is still used today.

The following equation (10.11 [46]) was derived from the Hertz equation and modified to include plastics materials properties and gear notation.

$$S_H = \sqrt{\frac{F}{fD_P} \frac{1}{\Pi \left[\frac{1-\nu_p^2}{E_p} \quad \frac{1-\nu_g^2}{E_g}\right] \frac{\cos\phi \sin\phi}{2} \quad \frac{m_g}{m_g+1}}} \tag{10.15}$$

$S_H$ = the Hertz contact stress

$F$ = tangential load

$D_P$ = the pitch diameter

$\nu_p$ and $\nu_g$ = Poisson's ratio for the pinion material and the gear material respectively

$E_P$ and $E_M$ = the modulus of elasticity for the pinion material and the gear material respectively

$\Phi$ = the pressure angle

$m$ = the speed ratio, $N_g/N_p$

$N_p$ and $N_g$ = the number of teeth of the pinion material and the gear material respectively

The surface contact stress calculated for the gear should be compared to a surface endurance limit for the material being used. This data is not usually available, but a comparison can be made to the compressive strength of the material to give a relative view of how the material might perform. The most reliable way to determine this type of data is through actual testing of gear sets running under the desired service conditions.

## 10.6 Gear Tolerances

For a gear to function properly, the tolerances that the gear is manufactured to are very important. Out of roundness, inaccurate tooth profiles and poor surface finish on the tooth face can cause excessive wear, noise and failure of the gear itself.

If the gear is to be produced by a molding process, it is imperative to properly account for the plastic shrinkage in the process. The tool has to allow for the diametrical shrinkage and also the shrinkage effect on the tooth profile. The molding process itself has to be precisely controlled to produce parts consistently with the same dimensions and tooth profiles. Machining can produce higher accuracy in the gear profile but would be considerably more expensive than a molded gear.

Since gears are always used in pairs, it is important to control how they mesh together. When two mating gears are brought together, the distance between their centers is half the sum of their pitch diameters, known as the standard center distance. For the gears to rotate at their standard center distances, the gears have to be perfect. Any errors in the dimensions, pitch angle or profile (tooth shape) of either gear will cause them to bind at some point as they rotate.

The pitch angle error plus profile error is called the tooth-to-tooth composite error (TTE). The amount the gear is out of round is called runout. Adding the runout to the TTE gives us the total composite error (TCE).

The standard method for checking gear accuracy is a center distance measuring device commonly called a gear checker. Simpler models have a dial indicator and the operator observes the displacement as the gear is rotated but, more common, is an electronic version that will plot and analyze results of the test. Figure 10.10 shows a schematic of a gear checker.

The gear to be measured (the test gear) is rotated in close mesh with a master gear of known accuracy. As the gear is rotated any deviations from ideal form or diameter will cause a radial displacement. The gear is rotated for one full revolution and the displacements can be charted as in Figure 10.11.

The chart shows the TCE generated one full revolution or the gear being measured. If the test gear and master gear were perfect, the result would be a straight line.

From the chart the accuracy or the gear can be evaluated and compared to the requirements of the AGMA (American Gear Manufacturers

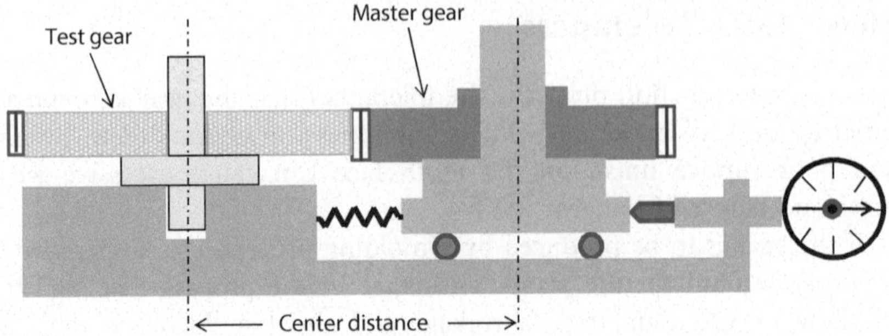


**Figure 10.10** Schematic of a gear checker.

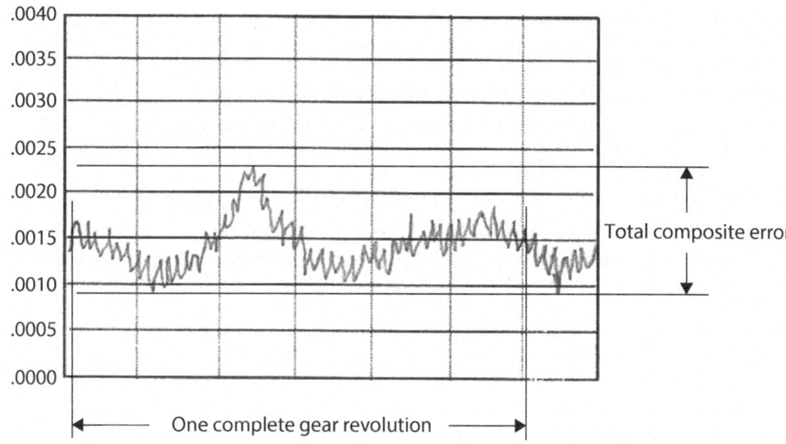


**Figure 10.11** Graph of total composite error.

Association), DIN (German Institute for Standardization), JGMA (Japanese Gear Manufacturers Association or ISO (International Standards Organization) specifications. Each of these standards will assign a quality number based on the TCE and TTE as well as the gear size. It should be noted that the AGMA has recently revised its standard to better align with the ISO standard. Prior to this change, AGMA had 13 quality classes designated Q3 through Q15 with 15 being the most precise. The new standard now has 10 classes with from A2 through A11 with the smaller number being the most precise, which is the convention used by the other referenced standards. The "A" designation is used to avoid confusion with the old "Q" numbers which will undoubtedly continue in use for a while. For reference, plastic gears molded to AGMA Q5-Q8 are common, Q9 and Q10 (approx. AGMA A7-A8 or ISO 7–8) [47] are more difficult and require a high degree of tool making and processing expertise. A very limited number of molders are just starting to achieve Q11 (approx. AGMA A6 or ISO6) levels. The gear tolerances at the higher levels exceed the tight tolerance level shown in Figure 5.3 of Chapter 5, General Design Considerations, and are very difficult to achieve.

## 10.7 Gear Tooth Design

When developing a plastic gear tooth profile design there are almost no limitations on what form can be machined into a mold. Any modifications that are deemed to be beneficial to the gear's performance can usually be incorporated. When choosing a profile design it is important for the designer to be sure to take into consideration how the plastic gear tooth will respond in actual operation.

Plastic gear teeth will deflect under load and in some cases the deflection can be considerable. In operation when a tooth deflects it puts the trailing tooth out of line for a truly smooth engagement with the next tooth from the mating gear. This can cause excessive wear, noise, and non-uniform motion of the gear set. To compensate for this deflection it is good practice to provide "tip relief." This is usually done by gradually thinning the tooth out starting from halfway up the addendum to the tip of the tooth. The amount required can be determined by calculating the amount the tooth will deflect under the design load.

As with any plastic part, sharp corners will cause stress concentrations that can cause premature failures and need to be avoided. At the bottom of the tooth form between each pair of teeth a full fillet radius should be incorporated. This can reduce the stress at this point by as much as 20% and will significantly improve gear life.

When designing two gears using a standard form, the smaller gear will have teeth that are thinner at the root than the larger gear. This will cause the smaller gear tooth to be weaker than the larger gear tooth and will limit the load that can be carried. To correct for this, the teeth and the smaller gear need to be thickened and the larger gear teeth thinned out to compensate for clearance until the roots of both teeth are the same thickness. This balanced tooth thickness will provide the optimum load carrying capability.

Two standard tooth forms that incorporate these changes are shown in Figures 10.12 and 10.13. Figure 10.12 shows the AGMA PT (American Gear Manufacturers Association) tooth form. Figure 10.13 shows the ISO (International Standards Organization) modified basic tooth rack. These two forms are commonly used for plastic gears and are essentially the same. The AGMA PT uses diametrical pitch and English units. The ISO rack uses the metric module, m, and metric units.

There are many other profiles available to choose from. For example, if you are designing gears for a control device, camera or watch you might want to choose a finer pitch form such as PGT-2 (Plastic Gear Technology, Inc.), PGT-3 or PGT-4. A number of software programs can help with this choice (see section on Software below).

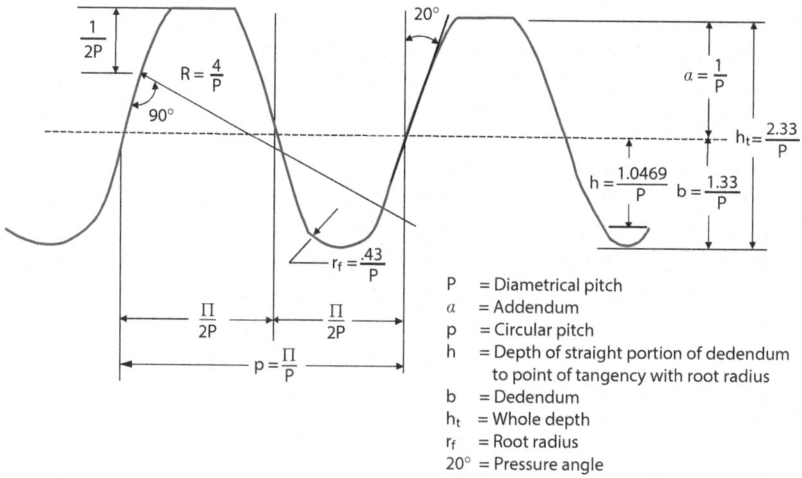


**Figure 10.12** AGMA PT tooth form.

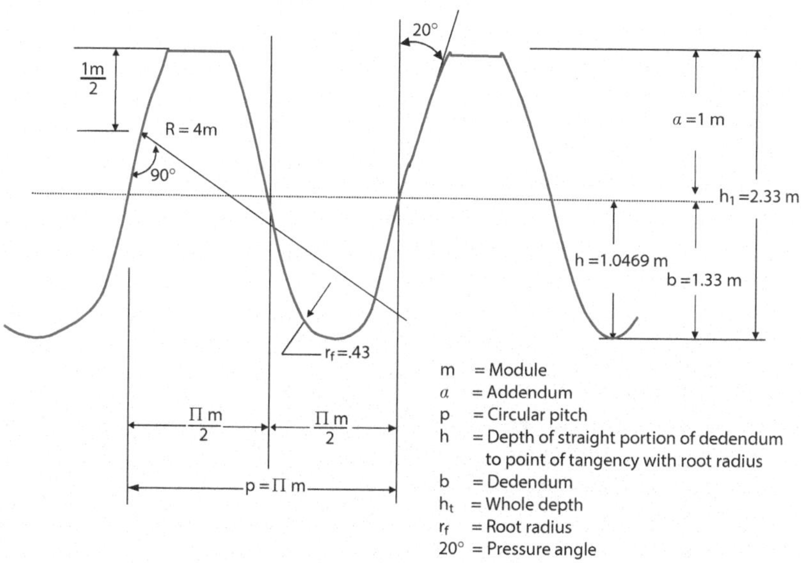


**Figure 10.13** Modified ISO basic rack.

## 10.8 Gear Mesh Conditions and Operating Distances

For gears to perform useful work they must be assembled with at least one other gear in some type of housing. When designing a gear, consideration must be given not only to the individual gear but also to its mate(s) and the components it will be assembled with.

When two gears are in mesh the distance between the two gear centers is half the sum of their two pitch diameters, which is called the standard center distance. In actual use this standard center distance can be affected by manufacturing tolerances, dimensional changes caused by changes in thermal expansion due to temperature changes and friction, dimensional changes due to moisture absorption, and runout in bearings used to mount the gears as well as similar changes in the mounting housing itself. If the center distance is too small, these changes could cause the gears to bind. A critical part of the gear design is to identify the minimum and maximum conditions and to adjust the center distance accordingly to compensate to ensure smooth running throughout the anticipated operating range. The equations that follow [48] will describe how to calculate these conditions. In the following equations in this section:

$\Delta C$ = change in operating center distance in center distance
$\alpha$ = coefficient of linear thermal expansion
$C_A$ = assembled center distance
$C_H$ = center to center distance on housing
$d_g$ = nominal operating pitch diameter of gear
$d_p$= nominal operating pitch diameter of pinion
H = gear housing
k = moisture (RH) coefficient of expansion
RH = relative humidity
$RO_{pb}$ = runout, pinion bearing or bushing
$RO_{gb,}$ = runout, gear bearing or bushing
$RP_{pb}$ = radial play, pinion bearing
$RP_{gb}$ = radial play, gear bearing
T = temperature
$TCT_p$, $TCT_g$ = total composite tolerance of pinion and gear respectively

First we need to look at the tolerances involved as we assemble the gears into place in some sort of housing. The center distance, C, between the holes or pins in the housing that the gears will be assembled in will have a tolerance from which can be determined $C_{min}$ and $C_{max}$. Additionally, the bushing or bearing (or if there isn't one, the hole or pin the gear will fit on) will have a runout tolerance. If a bearing is being used, there is also a radial tolerance to be considered. With this data, we can calculate the minimum assembled center distance as follows for an external gear:

$$C_{A\ \min} = C_{H\ \min} - \frac{RO_{pb} + RO_{gb}}{2} + \frac{RP_{pb\ \min} + RP_{gb\ \min}}{2} \tag{10.16}$$

To calculate the maximum assembled difference we also need to take into account the tolerances of the gears. For this we will start with the maximum housing center distance and also use the total composite tolerance (TCT) from the quality class of gear or whatever tolerance is specified:

$$C_{A\ \max} = C_{H\ \max} - \frac{RO_{pb} + RO_{gb}}{2} + \frac{RP_{pb\ \max} + RP_{gb\ \max}}{2} + TCT_p + TCT_g \tag{10.17}$$

For an internal gear we would modify the equations as follows. Note that for this case we subtract the composite tolerance in the $C_{\min}$ calculation and subtract the radial play in both calculations.

$$C_{A\ \min} = C_{H\ \min} - \frac{RO_{pb} + RO_{gb}}{2} - \frac{RP_{pb\ \max} + RP_{gb\max}}{2} \tag{10.18}$$

$$C_{A\ \max} = C_{H\ \max} - \frac{RO_{pb} + RO_{gb}}{2} + \frac{RP_{pb\ \max} + RP_{gb\max}}{2} - TCT_p + TCT_g \tag{10.19}$$

Next, we need to account for the effects that the operating environment of the gear assembly will have on the gears and the housing. This will include changes caused by temperatures, relative humidity (moisture) and exposure to chemical environments. During operation it is not unusual for the housing and the gears to see a different temperature than was present during assembly. This change will affect the dimensions of both the gears and the housing. We need to know the coefficients of thermal expansion, $\alpha$, of each material being used, as well as the temperatures they might see. Similarly, we will need the moisture coefficient of expansion, κ, for each of the materials involved, as well as the relative humidity (RH) range anticipated. Exposure to chemicals such as lubricants can cause plastic materials to change dimension. General quantitative data for this can be hard to find and it is highly advisable to do some testing to evaluate this effect.

For an external gear set we can calculate these effects using Equation 10.20 below:

$$\Delta C = \frac{d_g}{2}\left(\propto_h \Delta T_h - \propto_g \Delta T_g + (\kappa_H - \kappa_g)(\Delta RH)\right) + \frac{d_p}{2}\left(\propto_h \Delta T_h - \propto_p \Delta T_p + (\kappa_H - \kappa_p)(\Delta RH)\right) \quad (10.20)$$

If we are designing an internal gear set we would modify Equation10.20 as follows:

$$\Delta C = \frac{d_g}{2}\left(\propto_h \Delta T_h - \propto_g \Delta T_g + (\kappa_H - \kappa_g)(\Delta RH)\right) - \frac{d_p}{2}\left(\propto_h \Delta T_h - \propto_p \Delta T_p + (\kappa_H - \kappa_p)(\Delta RH)\right) \quad (10.21)$$

With the data from the above equations we can now calculate the maximum and minimum operating center distances, $C_{0\ min}$ and $C_{0\ max}$.

$$C_{0\ min} = C_{A\ min} + \Delta C \quad (10.22a)$$

$$C_{0\ max} = C_{A\ max} + \Delta C \quad (10.22b)$$

These values will define the full range of distances over which the gear set will be operated and should be used in gear geometry and load analyses.

## 10.9 Software

Over the years CAD software for plastic gear design has improved significantly, incorporating data generated from several decades of performance so it can optimize designs taking into consideration temperature, plastic shrinkage, moisture pickup and other environmental factors. This allows more accurate predictions of gear performance and shorter design cycles and time to market [49].

Many CAD platforms have gear design capabilities built in. In addition there are a number of add-on software suites available, including one from AGMA (https://www.agma.org/), KISSsoft (http://www.kisssoft.ch/english/home/index.php), GWJ Technology GmbH (http://www.gwj.de/en/home.html) and Dontyne Systems (http://www.dontynesystems.com/). Also, a number of plastics material suppliers have software that they will use when you are designing gears using their materials.

Most designs are now done on a CAD system, but it still is recommended practice to prototype and test whenever possible.

## 10.10 Prototyping

The importance of prototyping and testing to validate a gear design cannot be overemphasized. Particularly on a clean sheet design, achieving the perfect design on the first try is extremely difficult.

To provide the most useful information, prototyping should duplicate the production process as much as possible. If the gear is to be machined in production, obviously machining is the best. If the gear is to be molded, the same molding process should be used for the prototype as will be used in production. The single-cavity molds should be made of the production tool material and knockouts, cooling channels and venting should be as the production mold will be. It is extremely important to use the intended production gating system on the prototype mold as this will have a large bearing on the TTE to be achieved.

If 3D printing is used for the prototypes, testing will only be valid if 3D printing will be the production process. The gear dimensions and material properties will be different from a molded gear.

When testing, it is also important to test in accordance with how the gears will be used in operation. Gears that will operate under an intermittent high load should not be tested continuously, as this will generate a heat buildup that will not be seen in operation. With a low load gear, where heat buildup is negligible, a continuous test is possible once a negligible heat buildup is determined.

## 10.11 Gear Manufacturability

The tooth modifications we discussed above detail good design principles for the gear teeth, but in order to achieve the best quality gear of equal importance is the design of the gear body. The principles that were outlined in Chapter 5, General Design Considerations, are applicable here.

The first consideration should be the nominal wall thickness. It needs to be thick enough to provide the structural strength required and good plastic flow throughout the gear, but as thin as functionally possible to minimize shrinkage. For injection molded parts, depending on material, this thickness can commonly range from 0.030″ to 0.250″ (0.7 mm to 6.4 mm). It should be uniform throughout the gear if possible.

Figure 10.14 shows two possible configurations for a simple flat gear. In Figure 10.14A the gear is of uniform wall thickness throughout, with a center hole to mount the gear on a shaft. Typically, the gate will be located close to the center hole to achieve the most uniform flow of plastic to all of the gear teeth as possible. The material flowing from the gate to the teeth across from the hole will pack slightly less than the flow from the gate directly to the teeth on that side of the hole. Also, as the material flows around the hole and mates on the opposite side there will be a weld line where the two material flow fronts knit together. This can cause a slight distortion in the gear that will be observed in TTE testing and will cause an increase in the TTE. This can become an issue with larger gears and hard to flow materials. The situation can be improved by using multiple gates around the center hole, either with a hot runner or three plate mold. If the distortion just can't be tolerated the final solution would be to gate into the hole itself and drill the gate out with a secondary operation. Another possible design solution is shown in Figure 10.14B. If the gear shaft works only on one side and

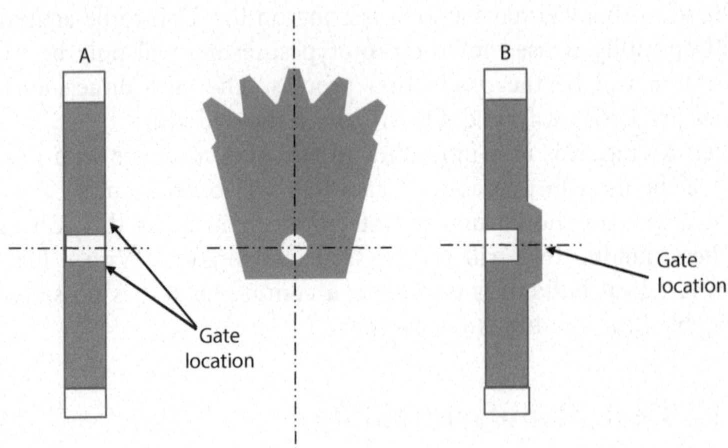


**Figure 10.14** Simple gear.

does not need to pass completely through the gear a blind hole can be designed to accept the shaft. The cover to the hole should be designed to provide uniform wall thickness to the gear body and the gate can be located directly in the center, which is the ideal gate location to insure uniform flow throughout the gear.

As the depth of the gear teeth increases it may be necessary to go to a hubbed or rimmed design to keep the molding walls at a reasonable thickness. Figure 10.15 shows some typical hub designs. In Figure 10.15A you can see the gear teeth attached to the hub with a thickness "h." The hub is connected to the gear hole by a disk of material that runs on one surface from the hub to the center hole. For adequate structural support you want the hub wall, "h," to be twice the thickness of the root of the gear tooth, "t" [50]. All of the plastic walls are the same thickness or slightly less than that of the hub. If for some other design requirement it is necessary to make the disc portion thicker, it could be done as long as the thickness around the periphery where it enters the hub is reduced to prevent sink in the adjacent areas of the gear teeth. Note also that all inside radii have very generous fillets. The radii should be at least ½ of the wall thickness and more is preferred (a ¼″ wall thickness should have a 1/8″ radius, minimum).

Design "A" can result in different shrinkage on the diameter on the disc side and the diameter away from the disc. Placing the connecting disc in the center of the gear equidistant from either side (10.15B) is a

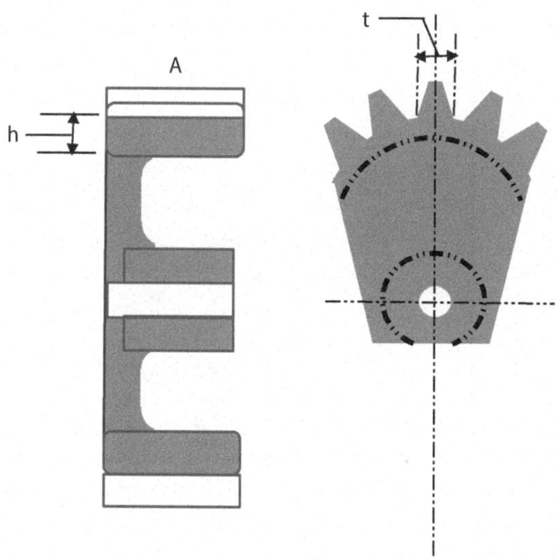


**Figure 10.15** Hubbed gear.

good approach to gaining more uniform shrinkage of the gear. Again, all inside fillets should be generously radiused. Gating issues will be the same as we had with the simple gear. The optimum quality will be obtained with a center gate. Adding holes through the disk to reduce material should be avoided as they will cause weld lines that can cause quality issues.

When using a hubbed design, particularly when the gear depth significantly exceeds the nominal wall thickness or the gear becomes large, it may become necessary to incorporate ribs into the design for structural support, stiffness or to improve the plastic flow to the gear teeth. Rib thickness should be no more than 50% of the nominal wall thickness with high shrink materials, such as unfilled nylon and acetal, and no more than 75–80% of the nominal wall for low shrink material like a filled PEEK. If possible, limit the rib height to no more than 3 times the thickness of the nominal wall. Deep ribs can be difficult to fill and as the depth increases, to keep within maximum wall thickness, less draft angle will be possible. This will make the ribs very difficult to eject from the tool. By using the centering design shown in Figure 10.15B, two shorter ribs can be used instead of a deeper rib for the same effect in design A. Where the rib intersects with the hub wall that is attached to the gear teeth we must deviate from our guidelines for fillets and restrict this fillet to no more than 25% of the nominal wall thickness. A larger fillet will cause a thick section that will cause extra shrinkage on the teeth next to each rib. When doing a TTE measurement of a ribbed gear, it is often obvious where each rib is located. Figure 10.16 shows

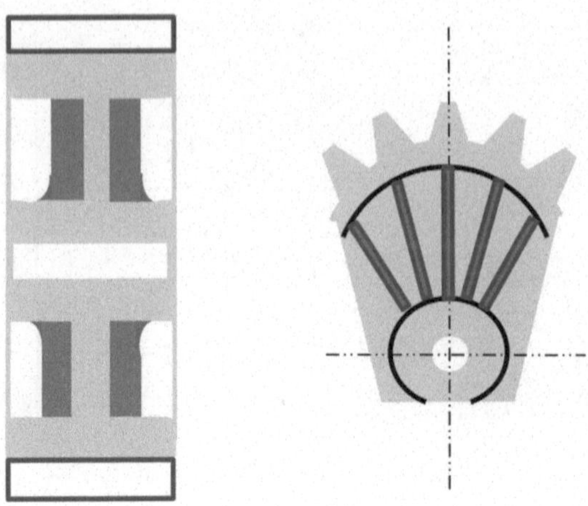

**Figure 10.16** Ribbed gear.

an example of a ribbed design. Ribs need to be spaced to leave adequate room for uniform cooling of the tool and part.

When designing a plastic gear there is the opportunity to incorporate other features into the gear such as another gear (cluster gears) cam, pulley or a snap fit. When doing this, the feature should be cored out as necessary to maintain uniform wall to insure the part will fill in the mold properly and the structural integrity and tolerances of the molded assembly is maintained.

## 10.12 Gear Materials

When selecting a plastic material for a gear application the designer needs to ensure that the material is strong enough to handle the gear tooth loading, dimensionally stable with adequate corrosion resistance to function in its intended operating environment and also have good wear and friction characteristics against the material to be used for the mating gear.

For wear resistance the most common test used is ASTM D3702: Thrust Washer Test, which evaluates wear performance and the dynamic coefficient of friction. In this test a sample disc of the material to be tested is pressed against a thrust washer (the contact surface). The thrust washer remains stationary while the sample is rotated on its surface under a set load pressure (P) and velocity (V). Testing at different PV conditions will determine a material's wear resistance capabilities. The volume of material lost is calculated by converting mass lost to volume (W). Dividing W by the testing force, velocity and elapsed time yields a wear factor K. This K value can be used to get a comparison of sliding wear resistance of one material over another. A lower value indicates a better wear resistance. The units for K typically are $in^3$ min/ft lb hr $\times 10^{10}$. Unfilled nylon 6/6 has a wear factor of 200, which is commonly used as the benchmark for comparing wear rates. Nylon 6/6 is used in many low load gear applications and a value over 200 would indicate a high wear rate. ASTM D3702 will also generate a dynamic coefficient of friction.

Another wear property of interest is the static coefficient of friction, which is the ratio of force required to cause motion between two contacting bodies and the force pressing them together tested by ASTM method D1894. This data is not always available for many plastics. Also, some material suppliers use modified or their own tests. When actually specifying a material, the specific data of interest should be obtained from the material manufacturer for the exact grade and properties desired. For a more detailed discussion of PV, coefficient of friction, and wear, see Chapter 13, Bearings.

Many plastic materials are suitable for specific gear applications. Nylon 6/6 was probably the first plastic material used for gears and still is in wide use. Other nylons used include 6/10 and 6/12. Acetal has found wide use, particularly in applications that are exposed to humidity and moisture such as washing machines and dishwashers. Polyphenylene sulfide and polybutylene terephthalates provide higher mechanicals and temperature resistance. Even higher temperature environments can be reached with thermoplastic polyimides and polyether ether ketones. Very precise, small, fine gears, like those used in watches, are molded from liquid crystal polymers. Ultrahigh molecular weight polyethylene is used where high impact is required. Thermosets are also used in some gear applications. Phenolics, usually filled with linen or canvas, are good for high temperature gears and will not burn. They are produced by compression molding and machining. Thermoset polyimides provide very high temperature service. Polyurethanes are used in applications where impact is important.

The suitability of plastic materials is further enhanced when they incorporate a number of performance additives. Polytetrafluoroethylene (PTFE) has the lowest coefficient of friction of any plastic material, and when incorporated in powdered form into a variety of plastics it significantly reduces the coefficient of friction of the compound. It will form a lubricating film on the part surface. Molybdenum disulfide works very well in applications of nylon running against metal. Silicon migrates to the part surface over time, providing a boundary layer. Graphite powder also provides good lubricity and works well in underwater applications. Adding these materials will lower some of the materials' mechanical properties. This can be offset by adding glass and carbon fiber reinforcements.

# 11

# Bearings

The purpose of a bearing is to minimize the friction and surface deterioration caused by surface contact due to sliding or rotating motion and pressure between two surfaces. Plastics unique properties, including inherent lubricity, toughness, wear resistance, chemical resistance and ease of processing, make it a very good choice for a number of applications. As in the other applications discussed, when designing bearings in plastic materials the designer also must take into consideration the variation of properties with temperature, moisture absorption, creep, thermal expansion and conductivity.

Plastic bearings are used in a wide range of applications, including shaft/housing applications in power tools, kitchen appliances, printers and copiers, water and chemical pumps, clocks and valve stems, as well as wear pads and ball joints.

When specifying a bearing material the tribological properties will be of primary interest. Tribology is the study of friction and wear of interactive surfaces that are in relative motion to each other. The specific properties of interest will be the coefficients of friction, wear and PV values.

The coefficient of friction, $\mu$, is illustrated in Figure 11.1. The coefficient of friction is equal to the ratio of the force opposing the forces applied to push one body along the surface of the other and the normal forces that are pressing the two bodies together.

There are two conditions which will create two types of friction coefficients. The first condition is when the two surfaces are stationary. The static coefficient of friction $\mu_s$ is the ratio of the amount of force required to initiate motion between the two surfaces and the normal forces pressing them together. The dynamic coefficient of friction $\mu_k$ is the ratio of the force required to keep the surfaces in motion and the normal forces pressing them together. While with most materials $\mu_s$ is usually greater than $\mu_k$, in most thermoplastics, $\mu_k$ is greater than $\mu_s$. Friction coefficients of plastics materials are typically in the range of 0.05 to 0.80. Coefficients of less than 0.2 are considered to be low. Dynamic coefficients of friction can vary with changes in pressure applied or surface velocity of the bearing relative to the material running against it.

Table 11.1 gives typical coefficients of friction of some selected plastic materials. Notice in the table that in addition to bearing material, the material that it is run against to determine the coefficient is given. As can be seen, the coefficient can vary significantly depending on what the bearing material is running against. Also, in general, a bearing material will not perform as well running against itself versus another material.

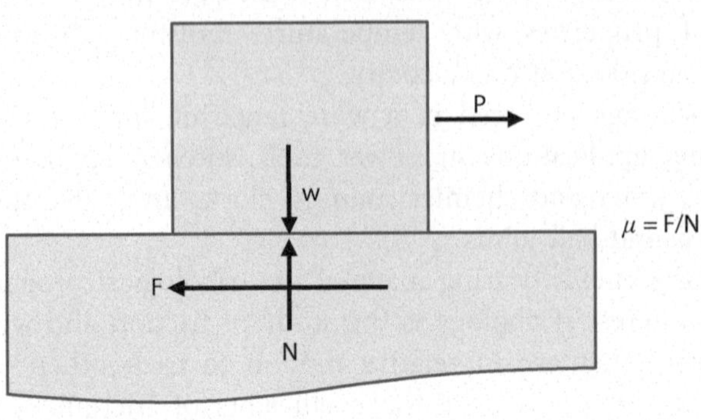


**Figure 11.1** Coefficient of friction between two bodies.

**Table 11.1** Typical coefficients of friction for some selected materials.

| Typical Coefficients of Friction, Selected Materials | | |
|---|---|---|
| acetal homopolymer on steel | 0.2 | 0.35 |
| acetal homopolymer on self | 0.3 | 0.4 |
| Nylon 6/6 on steel | 0.3–0.7 | 0.17–0.43 |
| Nylon 6/6 on self | 0.36–0.46 | 0.11–0.20 |
| Liquid Crystal Polymer on steel | 0.19 | 0.2 |
| Polyester on steel | 0.2–0.3 | 0.17–0.20 |

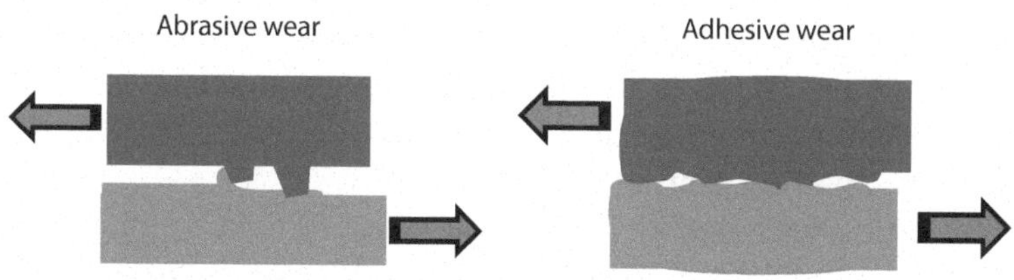


**Figure 11.2** Abrasive and adhesive wear.

## 11.1 Wear

Wear can be defined as the loss of material over time. Wear in plastic bearings is caused by pressure, temperature, relative surface speed and surface finish. There are two basic mechanisms of wear (Figure 11.2). The first, shown on the left-hand side of the figure, is abrasive wear. It is caused by the material running against the bearing being harder than the bearing material and as it rotates it abrades the softer bearing. It can be identified visually by distinct gouges, scratches or grooves present in the softer material. The second mechanism on the right is adhesive wear. In this, combinations of pressure, temperature and/or surface velocity will cause one material to stick and tear off, and transferring to the other material. When this happens, particles of one material can be seen adhered to the surface of the other.

As discussed in Chapter 10, Gears, wear resistance is measured by ASTM D3702, the Thrust Washer Wear Test. The test quantifies the amount of material a sample loses over time under prescribed conditions of pressure, speed and temperature. By rotating a sample of the bearing material under pressure against a thrust washer which can be made of the material intended to run against the bearing you can get a good simulation of the bearing material's long-term wear in the application.

The wear factor *K* is defined by Equation 11.1 below,

$$K = \frac{W}{F \times V \times T} \tag{11.1}$$

where *K* equals the wear factor measured in ($in^3$ min/ft lb hr) $\times$ $10^{10}$ or ($mm^3$/N-m) $\times$ $10^8$, *W* equals the volume wear in $in^3$ or mm, *F* equals the force in lbs or *N*, *V* equals the velocity in ft/min or m/sec and *T* equals the elapsed time in hr or sec. The lower the *K* value the better the wear resistance. *K* values are typically in a range of 1 to 1000. Unfilled nylon 6/6 with a *K* value of 200 is a benchmark to compare other materials to. A value in the 50–100 range would be classified as having excellent wear properties. Curves of wear factor vs temperature are useful in choosing bearing materials.

When testing with ASTM D3702, the pressure used in the test times the velocity used is known as the PV value. When comparing wear at a PV value, be sure that the conditions are the same. For example, a PV (pressure × velocity) of 2400 from a test at 40 psi and 60 ft/min might give very different wear data than the same PV run at 10 psi and 240 ft/min. Also, as PV values increase, the wear factors generally will also increase. Some manufacturers will provide curves of limiting PV values versus velocity for use as design guidelines. These charts will show what combinations of PV are within an acceptable design envelope. Figure 11.3 gives an example of a typical PV curve. Note that it gives the temperature at which the PV values are valid.

Let's look at an example that will illustrate how this data can be applied. We are designing a plastic journal bearing for a water pump. Figure 11.4 illustrates a typical journal bearing. In addition to minimizing friction and surface deterioration, a journal bearing also provides radial positioning for a shaft. The maximum load stress (W) is estimated to be 245 newtons. The shaft will turn at a maximum velocity of 180 RPM (N). The shaft will be 19 mm in diameter ($d_S$) and the bearing length (d) will be 30 mm long. Because the pump will be handling water, acetal copolymer has been identified as the bearing material of choice.

The PV value has to be calculated to determine if acetal copolymer will work in this application. First we have to calculate the pressure on the

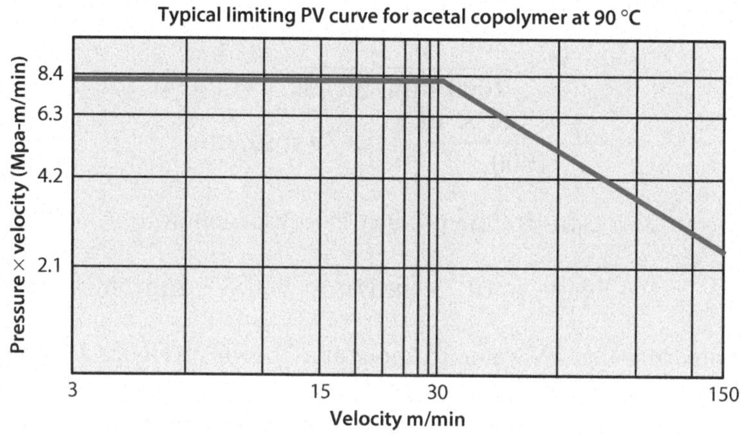


**Figure 11.3** Typical limiting PV curve for acetal copolymer at 90 °C.

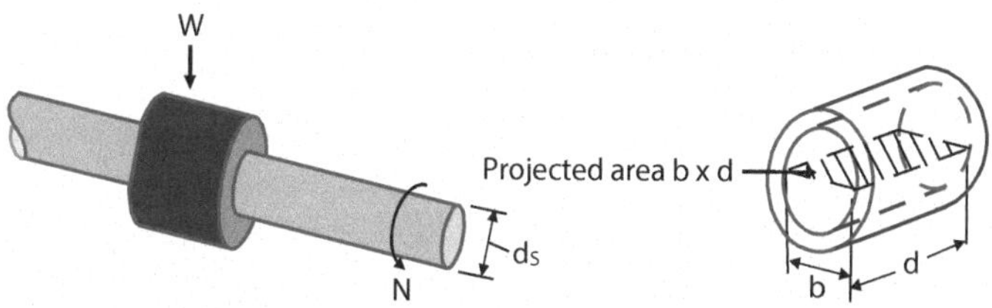


**Figure 11.4** Typical journal bearing.

bearing using Equation 11.2, where $W$ is the load on the bearing and the projected area the force acts upon is the width of the bearing, $b$, times the length of the bearing, $d$. For simplicity we assume that $b = d_s$ as any actual clearance will be negligible.

$$P = \frac{W}{(d \times b)} \tag{11.2}$$

Solving we find:

$$P = \frac{245}{(30 \times 19)} = 0.43 \text{ N/mm}^2 = 0.43 \text{ Mpa} \tag{11.3}$$

Next we need to find the surface velocity of the shaft on the bearing.

$$V = \frac{\Pi N\, d_s}{1000 \text{ mm/meter}} \tag{11.4}$$
$$V = \frac{3.14 \times 180 \times 19}{1000} = 10.75 \text{ mm/min}$$

We can now calculate the product P × V as follows:

$$\text{PV} = 0.43 \text{ Mpa} \times 10.75 \text{ mm/min} = 4.7 - \text{mm/min} \tag{11.5}$$

Comparing this PV value to the graph shown in Figure 11.3 we find that it is well below the allowable PV limit at a velocity of 10.75 mm/min. Also, the expected maximum operating temperature of 75 °C is well within the temperature limit of the allowable PV curve at 90 °C.

## 11.2 Bearing Life and Performance

When designing a bearing, the life and performance is dependent on a number of different factors. For journal bearings the hardness and surface finish of the shaft is very important. Centerless ground shafts with a hardness of $R_C$ 40 or greater are recommended. The surface finish of the shaft is more important than the surface finish of the bushing and should be on the order of 60 microns RMS or better. The rough spots on the softer bushing will tend to wear in, while rough spots on the shaft will tend to abrade the bushing. For critical applications polishing the shaft will improve wear. Care should be taken to polish the shaft in the direction the shaft will rotate. Polished steel generally will work better than chrome plating, as plating can have a porous surface that can be abrasive. Plastic bearings do not wear well with shafts of softer, non-ferrous metals such as aluminum, zinc or brass. If aluminum must be used it should be hardened or anodized.

Plastic bearings will usually require a greater running clearance than metal bearings due to the much higher coefficient of thermal expansion of most plastics as well as the tendency of some materials, such as nylon, to absorb water and swell. Also, molded bearings that run at elevated temperature can see some post-molded shrinkage over time. Bearing clearances normally will be in the range of 0.2% to 0.5% of the shaft diameter. Tighter tolerances can be used with some high performance bearing materials (see bearing materials discussion below). If you know the temperatures and conditions your bearing will be exposed to you can calculate the bearing clearance required from Equation 11.6 below [51].

$$ID_B = d(1 + \alpha S_M \Delta T_1 + C) + 2t\alpha S_B \Delta T_2 + 2t\kappa RH - D\alpha H \Delta T_3 \qquad (11.6)$$

Where
$ID_B$ = bearing ID
$D$ = bearing housing diameter
$d$ = shaft diameter at ambient temperature
$C$ = design (operating) clearance as % of shaft diameter
$\alpha S_M$ = thermal coefficient of expansion – shaft material
$\alpha S_B$ = thermal coefficient of expansion – bearing material
$\alpha H$ = thermal coefficient of expansion – bearing housing
$t$ = wall thickness of bearing
$T_1$ = temperature rise of the shaft
$T_2$ = temperature rise of the bearing
$T_3$ = temperature rise of the housing
$\kappa$ = moisture (RH) coefficient of expansion
$RH$ = relative humidity

The values for temperature rise can be determined through prototype testing or estimated from a similar application. If the housing is also made of plastic it might be necessary to add in changes in the housing due to changes in *RH*.

The accuracy of the bearing dimensions to the design intent is very important for the effective clearance. A molded bearing will likely have some out of roundness due to shrinkage in processing. This has to be accommodated in the design. This can be minimized by gate size, location and process conditions. Also, if a bearing is to be press fit into a housing, this can cause a reduction of the inside diameter of the bushing, tightening the clearance.

One of the advantages of a plastic bearing is the inherent lubricity of plastics materials which can be enhanced with the addition of friction reducing additives. While lubrication is not usually required with plastic bearings, pre-lubricating a bearing will facilitate the run-in period and significantly extend bearing life. It is especially important when it will be running against a soft metal.

Because of the clearance between the bearing and the material running against it, there is the possibility of external contamination getting into that space, particularly in a dirty operating environment. If operating conditions warrant, it is good practice to shield the clearance with an elastomer shield.

## 11.3 Bearing Design

The most common failure mode of plastic bearings is caused by overheating of the bearing. When designing a bearing the wall thickness should be held to a minimum to facilitate the heat transfer from the bearing to the housing it is mounted in. The following equation (11.7) gives a good empirical starting point for wall thickness [52], where $T$ equals the wall thickness in inches and $d$ equals the diameter of the shaft in inches.

$$T = 0.06d + .010 \tag{11.7}$$

Loading or impact requirements might necessitate a heavier wall. Also, with filled materials, a thicker wall may be necessary to provide for optimum mold flow to ensure complete fill and minimize weld lines.

Bearing lengths should typically be between ½ to 1–1/2 times the shaft diameter. If a longer bearing surface is required multiple bearings should be considered.

The outside of the bearing wall should be fully supported if possible to prevent deflection under load and also to provide a sink for heat generated during operation.

As a shaft runs in a bearing, some wear debris will likely be generated. This debris will cause additional adhesive or abrasive wear unless it can be removed from the bearing/shaft interface. A way to do this is to provide groove, holes or a combination of the two in the bearing surface to wipe any debris away from the bearing surface.

## 11.4 Bearing Materials

The generally low coefficients of friction make many plastic materials candidates for use as bearing materials. For very simple applications with low pressures and speeds, high density polyethylene will work well and is inexpensive. As the pressures increase, nylons and acetals provide good service. As temperatures and pressures increase polyesters and polycarbonates can be considered, and for more extreme applications, polyphenylene sulfides, poly(ether-ether keytones) and polyimides will provide superior service.

The performance of plastics bearing materials is commonly enhanced by the addition of a number of fillers. The most common is polytetrafluoroethylene (PTFE), which has the lowest coefficient of fiction of any

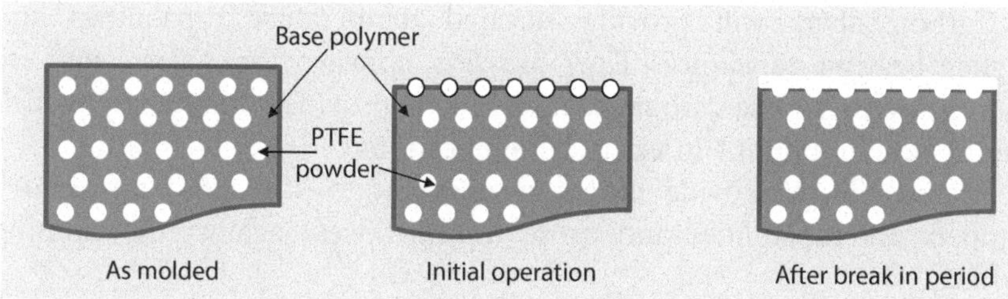


**Figure 11.5** PTFE lubricating mechanism.

plastic material. Adding 20% PTFE to a material can reduce the coefficient of friction by up to 60% and can reduce the wear factor by half or more. Figure 11.5 illustrates the lubricating mechanism of PTFE in a molded bearing [53]. When the bearing is initially molded the PTFE particles will be completely encapsulated by the base polymer. As the shaft or surface starts to operate against the bearing, it will wear some of the polymer away, exposing the PTFE. After further operation, the PTFE will "smear across the surface, forming a lubricating layer on the surface of the bearing. PTFE works well with almost any polymer but is relatively expensive.

Molybdenum disulfide works very well in applications of nylon running against metal. It acts as a nucleating agent for the nylon and creates a harder surface. It has an affinity for metals and provides a smoother metal surface. The material is black and will color the material it is added to Silicon (polydimethyl siloxane) migrates to the part surface over time providing a boundary layer. It is best in low load applications and is often used in combination with PTFE to provide initial friction reduction.

Graphite powder also provides good lubricity and works well in underwater applications. Again, its black color limits its applications to black or dark colors.

The addition of these materials can lower the mechanical properties of the base polymer. This can be offset by adding glass, carbon fiber or aramid reinforcements.

Glass fibers will improve a bearing's wear resistance and load bearing capabilities, but they will be abrasive on the material they are running against. Also, glass fibers can significantly change the anisotropic shrinkage of the composite material, which can affect the clearance. This can be offset somewhat by using a combination of glass beads with the glass fiber reinforcement.

Carbon fibers will provide elevated temperature capabilities and higher bearing capability. They are less abrasive than glass, but are more expensive. The carbon fibers can also provide conductivity, which can be very important in electrical applications.

Aramid fibers provide little improvement to strength but improve temperature capabilities and cause minimal wear against the running surface.

Material suppliers that provide bearing compounds provide a wide array of compounds with a variety of filler combinations that can meet most requirements. They always should be consulted for a material recommendation for any application.

# 12

# Pressure Vessels and Pipes

Plastic materials have been successfully used in a wide variety of pressure applications such as pipes, hoses, pumps, toilet flush valves, irrigation sprayers and butane lighters. As with other plastic structural components, as discussed in Chapter 6, the unique viscoelastic properties of plastic materials require taking into consideration not only the pressure involved but also the duration of the pressure, the operating temperatures, as well as other environmental conditions. The potential risks involved with a system under pressure make it imperative to be sure to use adequate safety factors.

That being said, the value proposition of plastics material and processing costs, as well as other advantages such as chemical and weathering resistance and lower relative weight, make it the material of choice for many applications. Advances in new engineering resins continue to provide more opportunities for additional applications.

## 12.1 Pipe

The simplest and most basic pressure vessel is a round pipe. Plastic pipe has been in commercial use for over 75 years. It is used in a wide array of applications, including water, gas and chemical process piping. Over 75 billion pounds of plastic piping is used globally, with PVC and HDPE being the largest players in the market.

For a thin-walled pipe or vessel the following formula can be used to determine the nominal hoop stress:

$$\sigma_H \frac{PR}{t} \tag{12.1}$$

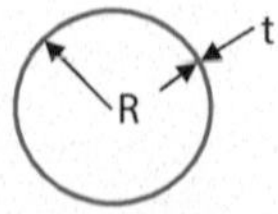


$\sigma_H$ equals the hoop stress, $P$ equals the internal pressure, and $t$ equals the wall thickness. This equation works quite well as long as the wall thickness $t$ < R/5. As the wall thickness becomes larger we need to use Equation 12.2 for stress in thick-walled tubes:

$$\sigma_H = P\,\frac{R_0^2+R_i^2}{t} \tag{12.2}$$

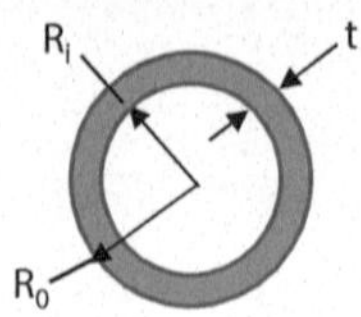


In this equation $R_0$ equals the outside radius of the pipe and $R_i$ equals the inside radius.

Using the stress calculated to select a material for performance will be valid if the pipe is pressurized just for a moment. If the pipe is to be pressurized over a longer period of time, we need to look at the effects of creep over long periods of time as well as creep at different operating temperatures.

When designing a pipe to carry water or gas, the intent generally is for that product to perform its function for years if not decades. Fortunately, we have a very large data base of performance data to draw upon. The plastic piping industry has developed an extensive database of pipe performance data.

The Plastic Pipe Institute (PPI) generates their data based on ASTM testing procedures. ASTM D-2837 outlines the procedure for obtaining the hydrostatic design basis (HDB) for thermoplastic pipe materials. Using test method ASTM D-1598, a series of pipe samples are pressurized under a range of pressures and the time-to-failure is recorded. When this series of failure stress-time points are plotted out on log-log paper, this results in a straight line that can be extrapolated out to 100,000 hours to determine the long-term hydrostatic strength (LTHS) of the material being tested [54]. This is analogous to the creep-rupture curves discussed in Chapter 6. From this value, ASTM D2837 assigns an HDB for the material. By multiplying the HDB times a service factor which considers variables and safety involved in a piping installation (a value of <1), a hydrostatic design stress (HDS) is arrived at, which is the estimated maximum tensile stress in the wall of the pipe due to internal hydrostatic pressure that can be applied continuously with a high degree of certainty that failure of the pipe will not occur. This HDS is the value used in design calculations. This data is temperature dependent and test data are available at a variety of temperatures. An extensive list of tested materials with their HDB and HDS values are available from the Plastics Pipe Institute in their publication TR4 [55].

The European manufacturers have developed similar data using International Organization for Standardization (ISO) test methods. ISO 9080 is similar to ASTM D-1598 and ISO 12162 is similar to ASTM D-2837. Instead of an HDB, the ISO method develops a minimum required strength (MRS) [56]. There are some differences between the data from each method to be aware of. The ASTM method extrapolates out to 100,000 hours to arrive at its HDB, the ISO method extrapolates out to 50 years. Instead of the service factor, F, the ISO standard uses a design coefficient C. While the MRS is typically lower than the HDB, when combined with the differences between F and C, the design stress for most materials comes out very close with either system. You just need to be careful which system you are working in.

Figure 12.1 shows a hydrostatic strength curve for PVC developed in accordance with ISO 12162 [57]. PVC-U is ISO terminology for unplasticized PVC or rigid PVC. The graph presents curves generated from hydrostatic testing at a series of temperatures starting at room temperature (20 °C) in increments of 10 °C. The curves are extrapolated out to 50 years (scale for years in bottom left corner). ISO 12162 gives a design correction factor of 1.6. To obtain a design stress, $\sigma_s$, divide the appropriate 50 year MRS value by the design correction factor.

Let's look at how we can apply this with an actual example. We want to move water at room temperature through a PVC pipe with a mean outside diameter of 73mm (2–1/2"). The pipe will be part of the water system in a house and will carry a municipal water pressure of 0.62 Mpa (90 psi). We

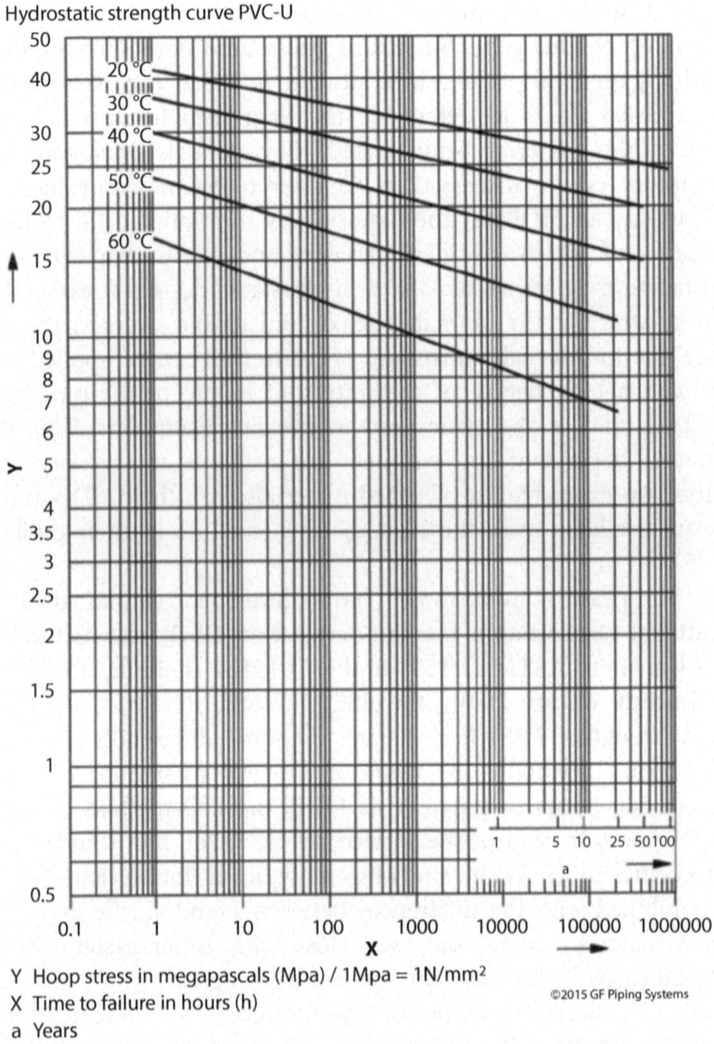


**Figure 12.1** Hydrostatic strength curve PVC-U.

want to have at least 50 years' service. What does the wall thickness of the pipe have to be designed to if we are to meet these conditions?

Looking at our hydrostatic strength curve for PVC we see that at 20 °C, we have a 50 year strength of about 23 Mpa. From ISO 12162 we find the PVC design correction factor to be 1.6. Dividing the *MRS* of 24 by our design correction factor we calculate the design stress $\sigma_S$ as follows:

$$\sigma_s = \frac{MRS}{C}\ \frac{23}{1.6} = 14.3 \text{ Mpa} \tag{12.3}$$

A useful formula for calculating pipe pressure ratings is the following ISO equation [58]:

$$P = 2 \times \sigma_S \times \frac{t}{D_m} \tag{12.4}$$

where $P$ is the pressure rating (Mpa), $\sigma_S$ is the design stress (MPA), $t$ is the wall thickness of the pipe and $D_m$ is the mean diameter of the pipe.

Rearranging Equation 12.4, we can now solve for $t$ to determine the wall thickness required to meet the specified service conditions.

$$t = \frac{P}{2 \times \sigma_S} \times D_m = \frac{0.62}{2 \times 14.3} \times 73 = 1.58 \text{ mm} \tag{12.5}$$

We find that a 73 mm pipe with a wall thickness or 1.58 mm will sustain a pressure of 0.62 Mpa for 50 years' service. It is important to recognize that this is based on a continuous steady pressure at a constant room (20 °C) temperature. If any upward excursions in temperature or pressure are anticipated, the wall thickness should be recalculated taking this into account.

## 12.2 Miner's Rule

Often, more typically than not, we find that an application does not run at a constant temperature or pressure over its lifetime. When faced with this situation, we can use an adaption of Miner's rule to calculate the effects. Miner's rule is widely used as a cumulative damage model for failures caused by fatigue. It is an empirically based procedure that is an approximation of the actual effect. An international standard, ISO 13760, has been established which "specifies a method for calculating the maximum allowable hoop stress applicable to pipes exposed to varying internal pressures and/or temperatures during their expected lifetime"[59].

For this topic the terms of the following equations are defined as follows:

$a_i$ = the fraction of a year, expressed as a percentage, under the set of "i" conditions

$t_i$ = the lifetime under a specified set of conditions "i" (I will equal condition 1, 2, 3, etc.,) expressed in years

$T_x$ = maximum permissible time of use, under varying conditions, expressed in years

$TYD$ = Total yearly damage, expressed in years

We can calculate the total yearly damage, *TYD*, expressed as a percentage of total permissible damage using Equation 12.6 below.

$$TYD = \sum \frac{a_i}{t_i} \tag{12.6}$$

We can calculate the maximum permissible time of use, $T_x$, in years using Equation 12.7.

$$T_x = \frac{100}{TYD} \tag{12.7}$$

We can work through an example which will better illustrate the application of these equations to a practical problem. A PVDF piping system is being used in a pharmaceutical industry application [60]. The pipe will be handling purified water at 25 °C and 5 bar pressure. The system has to be sterilized daily and will be subjected to saturated steam at 135 °C and 2.2 bar pressure for 30 minutes every day. The facility specifications require an operating lifetime of 15 years. We need to determine whether the pipe and material selected will meet this requirement. We have two sets of conditions, $t_i$, to evaluate. In the first, $t_1$, the lifetime of the pipe with water at 20 °C and 5 bar can be found from the hydrostatic curves to be 50 years. For the second condition, $t_2$, at 135 °C and 2.2 bar the lifetime is found to be 0.51 years. We find the fraction of the year, $a_i$, at each set of conditions as the time under the first set of conditions, $a_1$, to be 97.9 percent and the time at the second set of conditions, $a_2$, to be 2.1 percent. We now have the information we need to solve the problem. We first combine Equations 12.6 and 12.7 and format to evaluate the two sets of conditions as follows:

$$T_x = \frac{100}{\frac{a_1}{t_1} + \frac{a_2}{t_2}} \tag{12.8}$$

Substituting into Equation 12.8 we solve for $T_x$:

$$T_x = \frac{100}{\frac{97.9}{50} + \frac{2.1}{0.51}} = 16.5 \text{ years} \tag{12.9}$$

We find the maximum permissible use time for this combination of conditions is 16.5 years, which exceeds the desired life of 15 years.

Plastic pipe fittings are a little more complex than plastic pipe for several reasons. The first are the issues of geometry. There are many different configurations of pipe fittings ranging from a coupling, that is very similar to a pipe, to tees and elbows that have geometrical features that cause stress concentrations which will limit the fitting hydrostatic performance. The manufacturing process for fittings also generally leads to higher amounts of stress than that found in a pipe. Although there is some long-term hydrostatic data available for molded pipe fittings material, it is much more limited than that for pipe. A common practice to accommodate these issues is to set minimum fitting wall thickness at 125% of corresponding size pipe wall and confirm adequacy through product testing. This works well for expanding a similar product line where there is considerable empirical data to support it. Designs could also be calculated individually as discussed below.

## 12.3 Other Pressure Vessels

When designing pressure vessels, one has to be extremely cautious about safety. Plastics under pressure can fail catastrophically with incorrect design or manufacturing defects. Adequate safety factors such as recommendations from the ASME (American Society of Mechanical Engineers) or those established from similar designs in the past are a good starting point. With pressure vessels, the design is not complete until prototypes can be made and tested under laboratory conditions and the design evaluated and modified as required.

Most pressure vessels are essentially cylindrical or elliptical pipes that are capped at both ends. The geometry of the end configuration can cause areas of high stress concentration and will control the pressure carrying capability of the pressure vessel. These areas can be examined by finite element analysis (FEA) programs to determine stress levels and help establish safety factors. Also, it is important to account for any normal variability in the manufacturing process.

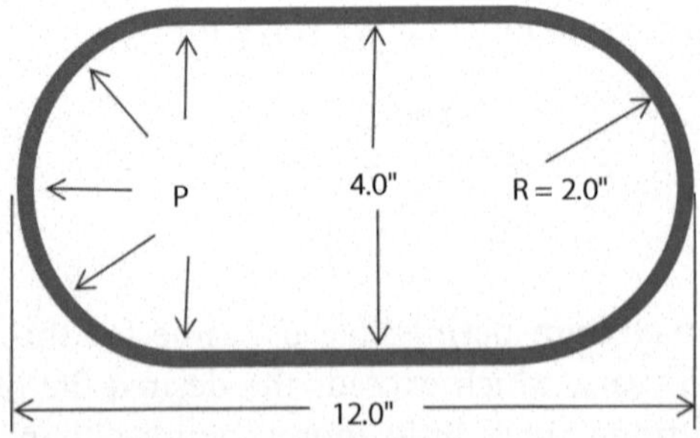


**Figure 12.2** Cylindrical pressure vessel example.

Let's look at an example to illustrate the design process. Figure 12.2 shows a cylindrical pressure vessel with hemispherical ends. The inside diameter is specified at 4.0 inches and the length to be at least 10" to accommodate the desired volume. The pressure fluid will be chlorinated water. The operating pressure will be an intermittent 40 psi and the desired burst pressure for the vessel is 350 psi. The operating temperature will be 730 °F in a laboratory environment. An acetal copolymer has been the initial material selection because of its resistance to the chlorinated water, good impact strength and ductility. We need to find the wall thickness of the vessel and validate the material choice.

The first step is to determine the tentative wall thickness using Equation 12.1 for hoop stress in thin-walled tubes.

$$\sigma_H = \frac{PR}{t} \tag{12.1}$$

Based on our finite element analysis or other pertinent information we can find what the maximum stress will be using Equation 12.10 below, where $\sigma_{max}$ equals the maximum stress, $\sigma_H$ equals the nominal hoop stress, and K equals the stress concentration factor.

$$\sigma_{max} = K\sigma_H \tag{12.10}$$

The maximum stress concentration will occur where the hemisphere ends intersect with the cylindrical tube. There could also be an area of stress concentration at the inlet or outlet to the vessel, which we have omitted to simplify this example, and also process conditions in manufacture, which we will discuss a little later on, can affect the amount and location of stress. Based on an FEA and previous experience, we set

$K$ equal to 3. We find that the maximum allowable stress for our unfilled acetal copolymer is 8800 psi [61].

By substituting Equation 12.10 into Equation 12.1 and rearranging to solve for thickness we get:

$$t = \frac{KPR}{\sigma_{\max}} \tag{12.11}$$

Substituting and solving we find that:

$$t = \frac{3 \times 40 \times 2}{8800} = 0.239'' \tag{12.12}$$

We calculate the wall to be 0.239″ . Just a quick check to be sure it was OK to use the thin wall pipe formula we find R/5 to be 0.40", which is greater than the calculated wall, so the equation is valid for this problem.

The wall thickness we calculated is designed to meet the specified burst (short time failure) pressure that was specified. We now need to validate that this design will function with the operating pressure desired. We can use an ASME guideline that postulates that operating pressure should be less than 1/6 of the short-term burst pressure, or, to express in an equation [62]:

$$P_{op} = \frac{\mathrm{P}_{\mathrm{burst}}}{6} \tag{12.13}$$

Solving we find that:

$$P_{op} = \frac{350}{6} = 58.3 \text{ psi} \tag{12.14}$$

This is greater than our required operating pressure of 40 psi, so this design meets the requirements. We can calculate the safety factor for the planned operating pressure as follows [58]:

$$\sigma_{\max\ \mathrm{op}} = \frac{KPR}{t} = \frac{40 \times 2 \times 3}{0.239} = 1004.2\text{psi} \tag{12.15}$$

And the safety factor is:

$$\text{safety factor} = \frac{\sigma_{\text{max}}}{\sigma_{\text{max op}}} = \frac{8800}{1004.2} = 8.7 \qquad (12.16)$$

The next step that must be performed is to obtain a prototype or initial production sample and perform burst tests and cyclic pressure tests to validate the design assumptions.

In Chapter 6 we discussed the mechanism of creep, where plastics will continue to deform over time when subjected to a constant load. When we design plastic parts which must function under load for a period of time, instantaneous strength data is no longer adequate for our calculations. In the design of pressure vessels that will be under pressure for a period of time we need to incorporate long-term creep data into our calculations. This is particularly important to ensure that adequate safety has been built into the design parameters.

We will look at another example to illustrate what data is needed and how it is applied. We need to determine the wall thickness that will be required for a cylindrical pressure vessel similar in shape to the one in Figure 12.2 but with a diameter of 75 mm and will operate at a continuous pressure of 0.28 Mpa and a temperature of 65 °C. It is required that the vessel operates for a minimum of 10 years.

In order to solve this problem we need to determine what the design hoop stress of the material that is chosen is at 10,000 hours and 65 °C. We can get this from the creep rupture curve.

Based on the temperature requirement we can quickly rule out the unfilled acetal we used in the previous example. After looking at several possibilities, a nylon 6/6 is chosen. Figure 12.3 shows the creep rupture curve for this material at the required conditions. Looking out at the 10 year point on the x axis we find the hoop stress at that point to be 18 Mpa. Again, we need to apply a safety factor to account for stress concentration, which we can develop by FEA or get from other sources. For this example we will choose a factor of 3.

To find the wall thickness we will use the same equation we previously used that we derived from substituting Equation 12.10 into Equation 12.11 and rearranging it.

$$t = \frac{KPR}{\sigma_{\text{max}}} \qquad (12.17)$$

Substituting and solving we find that:

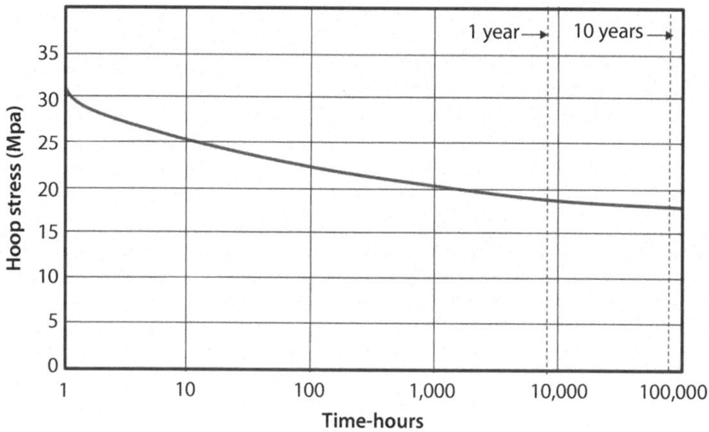


**Figure 12.3** Typical hoop stress vs time-to-failure nylon 6/6 @ 65 °C.

$$t = \frac{3 \times 0.28 \times 37.5}{18} = 1.75 \text{ mm} \tag{12.18}$$

## 12.4 Other Types of Pressure Vessels

The ability of plastics to be manufactured in almost any shape makes it possible to design a wide variety of differently shaped pressure vessels besides the simple cylindrical shapes we have discussed. Figure 12.4 gives stress formulas for a few of the more common types [63].

## 12.5 Material and Manufacturing Considerations

When choosing a material for a pressure vessel, materials that maintain their strength when loaded over time are very desirable. This can lead you to selecting a filled material, but you must do so with caution. As discussed in Chapter 3, Overview of Plastics Materials, fiber reinforced plastics will have orthotropic or anisotropic properties; properties such as shrinkage, tensile strength, modulus of elasticity and coefficient of thermal expansion that is different in the flow direction and in the transverse direction. Analyzing these materials can only be done through finite element analysis. Material suppliers should be consulted for the suitability of any material in pressure applications.

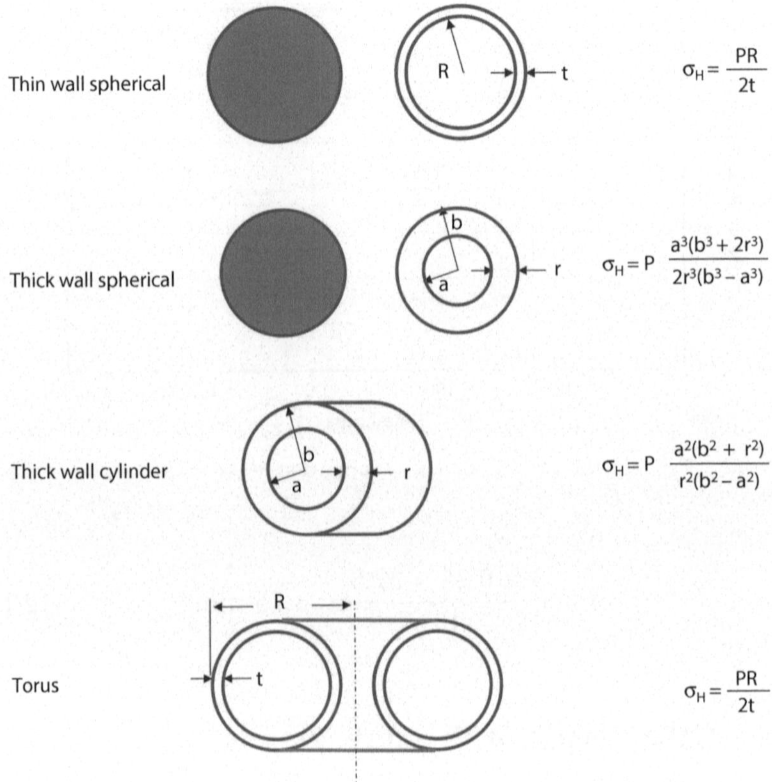


**Figure 12.4** Stress formulas for some common pressure vessel shapes.

Pressure vessels can be manufactured with most plastic processes but the most common are injection molding, blow molding, extrusion, and filament winding to create vessels reinforced with continuous fibers.

The single most important factor to be considered is the minimization of stress and stress concentrations. Wall thicknesses need to be as uniform as possible and any changes of direction need to be transitioned with generous radii. In molded parts, gate size and location are critical to maximize uniform flow and minimize molded in stresses. Mold flow software is very helpful in this regard. Weld lines need to be avoided or minimized, as they will likely be a source of failure.

When designing a vessel, the shape of the ends is very important. Flat ends are the worst configuration, with elliptical and round ends each being progressively better. The round and elliptical ends don't work if the intent is to stand the cylinder on its end. This can be handled by molding a base extending from the vessel for it to sit upon

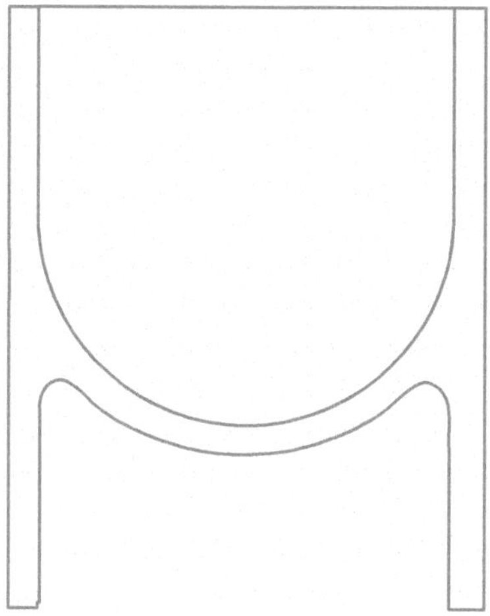

**Figure 12.5** Skirt base.

like in the example in Figure 12.5. Care needs to be taken to minimize increases in wall thickness where the base meets the body of the pressure vessel.

Quality control of parts intended for use under pressure should be adequate to ensure that the part design will meet the design criteria and also ensure that the manufacturing process is consistently under control. Specific industries, such as pipe and fittings, have standards developed that can be used to help develop a quality plan.

# 13

# Plastic Optics

Many amorphous plastic materials possess a combination of properties that make them an excellent choice for a number of optical applications such as light pipes, mirrors, lenses, and compact discs. Plastic materials can exhibit high coefficients of light transmission (the % of light that passes through a material), can be produced with high quality surfaces and complex shapes and can be manufactured very economically. This allows the designer to explore highly functional designs with a great deal of latitude. Table 13.1 gives some typical examples of light transmission values for plastic materials commonly used for optical applications, as well as some comparative materials.

## 13.1 Optical Fundamentals

Let's look at some fundamental optical properties that affect the design of optical components. The first is that of reflection. When a ray of light strikes a planer surface, some of the light may bounce off of the surface. This is called reflection. As can be seen from Figure 13.1, the angle that the light strikes the surface, called the angle of incidence, is

**Table 13.1** Light transmission.

| Material | Transmission |
|---|---|
| Polymethyl methacrylate | 92% |
| Polystyrene | 90% |
| Polycarbonate | 88%–91% |
| Glass | 92% |
| Copolyester | 85%–88% |

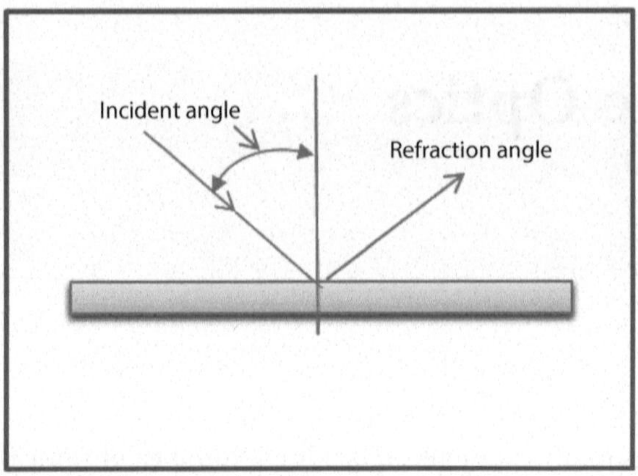


**Figure 13.1** Light reflection at a surface.

the same as the angle of reflection, the angle of the light as it leaves the surface.

If the material that composes the surface is not opaque and has a different refractive index from the adjacent material the light is coming through, some of the light may be reflected, but some may also pass into the second material and be refracted. Refraction is the bending of the path of a light wave as it passes from one material into another. The refraction occurs at the boundary between the two materials and is caused by the change of the speed of light as it crosses the boundary. The speed of light is dependent on the optical density or index of refraction of the material it is passing through. Table 13.2 below gives the index of refraction of a number of common optical materials.

**Table 13.2** Index of refraction.

| | |
|---|---|
| Vacuum | 1.000 |
| Air | 1.0001 |
| Water | 1.33 |
| Polymethyl methacrylate | 1.491 |
| Polystrene | 1.59 |
| Polycarbonate | 1.58 |
| Glass | 1.19–1.50 |
| Cyclic Polyolefins (COC) | 1.53 |
| Copolyster | 1.545 |
| Diamond | 2.42 |

The index of refraction is a measure of how a light ray will deflect or bend (the angle of refraction) when it moves through one material into another.

We can calculate the angle of refraction using Snell's law, which is $n_1\sin\Phi_1 = n_2 \sin\Phi_2$, where $n_1$ is the index of refraction of material 1, $\Phi_1$ is the incident angle, $n_2$ is the index of refraction of material 2, and $\Phi_2$ is the refraction angle. Figure 13.2 demonstrates this principle.

The difference between the index of refraction for air and a plastic material is responsible for the ability of plastics to pipe light, which will be discussed further below.

When light travels in a denser medium, such as a transparent plastic, and comes to an interface with a less dense material, such as air, the light ray will refract and remain in the denser material as long as the angle of incidence is less than a critical angle which can be calculated and is dependent on the refractive indices of the two materials. A higher refractive index will reflect a wider angular range of light (Figure 13.3). The amount of light that will be reflected internally can be calculated using the Fresnell equation:

$$R_F = \frac{N_2\cos\theta_1 - N_1\cos\theta}{N_2\cos\theta_1 + N_1\cos\theta} \tag{13.1}$$

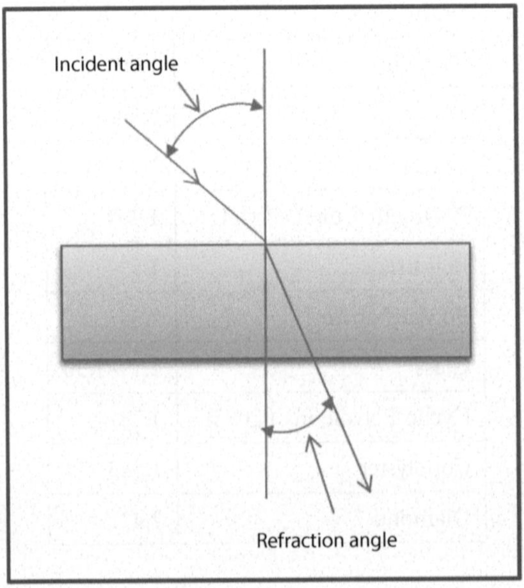


**Figure 13.2** Angle of refraction.

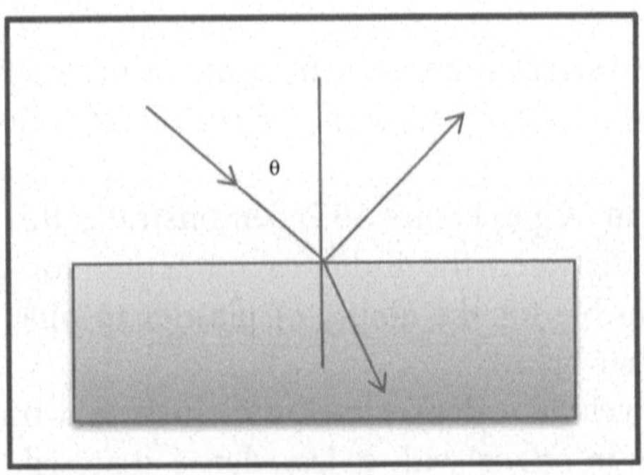


**Figure 13.3** Refracted light and critical angle.

The direction a light wave will bend when it passes from one material into another depends on whether the light speeds up or slows down as it crosses the boundary. If you draw a normal line perpendicular to the boundary surface, if the light slows, the ray will bend toward the normal. If the light speeds up, the ray will bend away from the normal. In Figure 13.4, the incident ray is passing through air (index of refraction 1.0001) and entering a polymethylmethacrylate lens (index of refraction of 1.491). The light ray will slow and the refracted ray will bend towards the normal line.

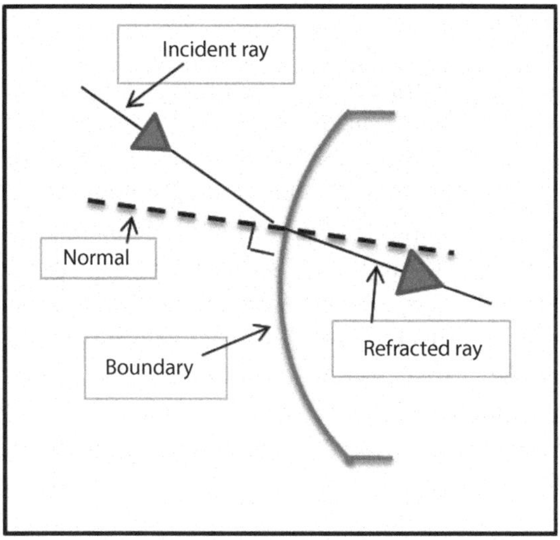


**Figure 13.4** Bending of light between different refractive indices.

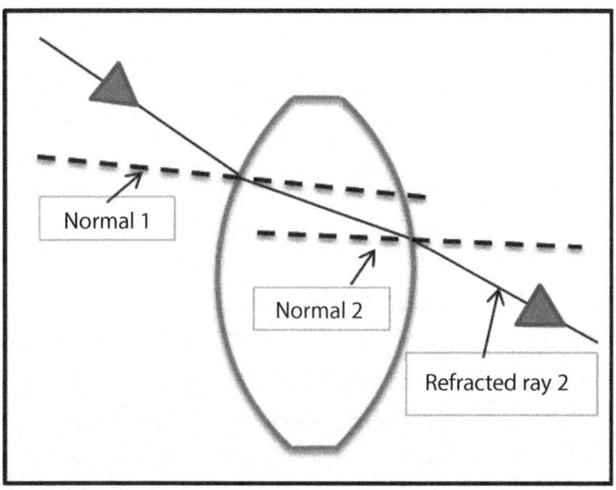


**Figure 13.5** Example of two refractions.

In Figure 13.5 we see what happens as the light ray passes through the lens and back into the air. Again, the light ray is refracted, but this time it bends away from the normal line drawn perpendicular to the boundary at the point the light ray crosses it.

## 13.2 Mirrors

Now let's look at some applications of these principals. The first we will discuss is mirrors. Mirrors can be made in two basic ways, first surface or second surface mirrors. This refers to which surface of the mirror the light is reflected off of, as shown in Figure 13.4. We make a mirror by covering a surface, usually by vacuum metallizing or electroplating with a reflective opaque reflective surface (special applications can be made with a semi-opaque surface that allows some of the light to pass through and not be reflected). With a first surface mirror, the light does not enter into the mirror material, and therefore, any plastic material which can be plated or metallized can be used. Platable grades of ABS are often used for first surface mirrors. With a second surface mirror, the light enters the mirror through the first surface and is reflected off the back surface, so obviously, the mirror itself must be transparent. Acrylics and polycarbonates are commonly used as materials for second surface lenses. With first surface mirrors, the reflective coating is on the outside surface and can be susceptible to scratches and damage. Second surface mirrors are usually easier to protect in a design application.

Mirrors can be made as plano mirrors, which reflect the image back as presented by making the mirror thickness uniform and the surfaces flat. They are usually made by injection molding and by casting. This sounds easier than it is, as making a truly flat surface without distortion, depending on how it is made, requires good processing techniques and tooling.

Mirrors that have a concave spherical surface will concentrate or reduce the reflection of an object. Mirrors can also be made to magnify, as in a cosmetic mirror, by making a convex curved mirror surface which will enlarge the reflected surface.

Figure 13.6 is a schematic of a concave mirror system. The distance of the object to be reflected to the center of the mirror surface is the object distance O. The distance from the center of the mirror surface to the place the image is formed is called the image distance I. The radius of the mirror is R. For a concave mirror, as shown, the radius of the mirror is positive (+). If the mirror was a convex mirror the radius of the the mirror would be considered to be negative (−).

The relation of the object distance, the image distance and the radius can be described by Equation 13.2.

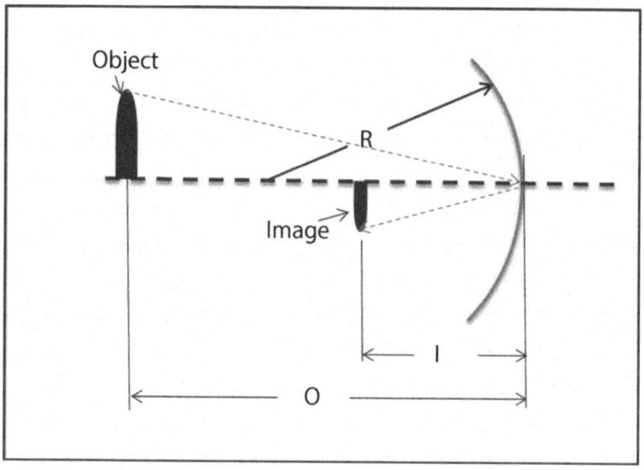


**Figure 13.6** Concave mirror system.

$$\frac{1}{0} + \frac{1}{I} = \frac{2}{R} \tag{13.2}$$

This allows us to determine the radius we require for a mirror when we know where our object will be and where we would like the image to be. For example, if we are designing a mirror for an automobile headlight, we know the filament or the light is going to be 4 inches in front of the mirror. Our specification requires that the image forms 10 feet (120") in front of the light. The radius that is required can be calculated from the equation above as follows:

$$\frac{1}{4} + \frac{1}{120} = \frac{2}{R} \qquad R = 7.74'' \tag{13.3}$$

Since the radius we calculated is positive, we know that we will need to use a concave mirror.

The magnification of a mirror can be calculated by the formula [64]:

$$m = -\frac{I}{O} \tag{13.4}$$

where $m$ equals the magnification ratio and $I$ and $O$ equal the image distance and the object distance respectively.

## 13.3 Light Pipes

Light pipes utilize the internal reflection that occurs at an interface, such as plastic and air, to keep the light within the plastic pipe and transmit light from a source through the pipe to an outlet.

The largest angle of incidence at which refraction can still occur is called the critical angle. This occurs when the refracted ray at the boundary has an angle of refraction of 90 degrees. For any angle of incidence greater than the critical angle, the light will undergo total internal reflection. We can derive a calculation for the critical angle from Snell's law and find the critical angle $\Theta_{\text{crit}}$ as follows:

$$\Theta_{\text{crit}} = \text{sine}^{-1}\left(n_r/n_j\right) \tag{13.5}$$

where $n_i$ is the index of refraction of the incident material and $n_r$ is the index of refraction of the refractive material. For polycarbonate refractive index 1.58 and air (refractive index 1.0001)

$$\begin{aligned} \Theta_{crit} &= \text{sine}^{-1}\ (1.001/1.58) \\ &= \text{sine}^{-1}\ (0.634) \\ &= 39^\circ \end{aligned}$$

## 13.4 Lenses

A lens has two refracting surfaces. The image formed by the first service serves as the object for the second surface. Figure 13.7 is a schematic of a simple lens system. The system can be defined mathematically by the equation below:

$$\frac{1}{O} + \frac{1}{I} = (n-1)\left[\frac{1}{R_1} - \frac{1}{R_2} + \frac{(n-1)t}{nR_1R_2}\right] \tag{13.6}$$

where $O$ is the distance of the object to the lens, $I$ is the distance of the image formed from the lens, $n$ is the index of refraction of the lens material, $R_1$ and $R_2$ are the radii of lens surfaces 1 and 2 respectively, and $t$ is the thickness of the lens measured along the central axis. This is known as the Lensmaker's equation.

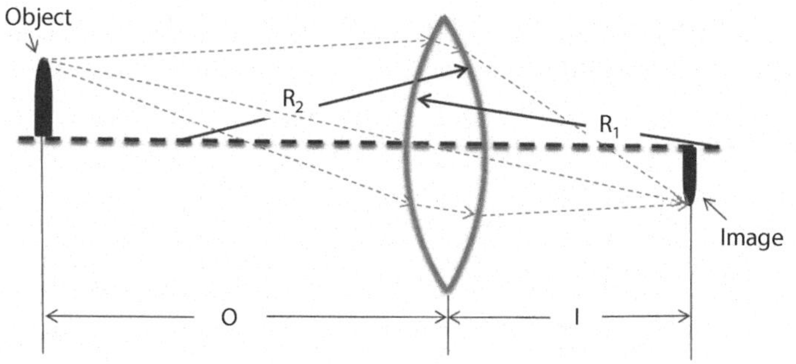


**Figure 13.7** Simple lens system.

If the lens is thin and $t$ is negligible compared to $R_1$ and $R_{2,}$ the equation can be simplified to the thin lens formula shown below:

$$\frac{1}{O}+\frac{1}{I}=(n-1)\left[\frac{1}{R_1}-\frac{1}{R_2}\right] \tag{13.7}$$

This simplified formula is very useful to provide quick assessments of a general lens configuration. Also,

$$\frac{1}{f}=\frac{1}{0}+\frac{1}{I} \tag{13.8}$$

where $f$ equals the focal length of the lens. The focal length will be positive for a converging lens and negative for a diverging lens. If we know any two of the object distance, image distance and focal length, we can easily solve for the third.

For example, a candle is placed 20 inches away from a lens with a focal length of 8 inches. Using the equation above we can find the image distance as follows:

$$\frac{1}{8''}=\frac{1}{20''}+\frac{1}{I} \qquad \frac{1}{I}=0.075'' \qquad I=13.33'' \tag{13.9}$$

Actual design and optimization of lens systems can be quite complex with many variables and are usually done with optical design software packages such as OSLO (Optical System and Lens Optimization) from Lambda Research Corporation.

## 13.5 Manufacturing Processes for Optical Components

Optical components are especially demanding to manufacture because they can require extremely tight tolerances. Conventional techniques can be used, such as injection molding, compression molding and extrusion, but there also are a couple of special techniques that are used that will be discussed here.

The first is injection-compression molding, also sometimes called "coining." The coining process gives the advantages of combining the good melt preparation and rapid cycling of injection molding with the tight tolerance capability and low stress parts of compression molding. The process uses a conventional injection molding machine to prepare the melt and inject it into the mold, but either through a machine modification or in the mold design itself; after the plastic is injected into the cavity the mold will close or compress, applying a uniform pressure to the cavity to fill out the part that will essentially be stress free. A schematic of the process is shown in Figure 13.8. After the plastic is injected the right-hand side of the tool is moved to the left, uniformly compressing the material in the cavity. The low stresses reduce the amount of warpage and variations in shrinkage across the part. The disadvantages of this process are that it requires an extra closing mechanism on either the molding machine or the mold, which adds to the cost and the process complexity, and the movement of the compression somewhat limits geometric configurations that are reasonably possible.

Another technique that is commonly used to produce high quality optics is liquid resin casting. In liquid resin casting acrylic prepolymer

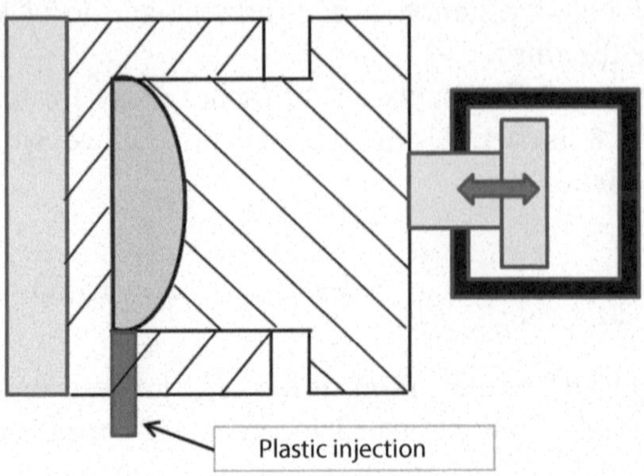


**Figure 13.8** Coining.

syrup is injected under low pressure into a mold, where it is cured. This is the process used for plastic eyeglass lenses. The mold is usually made up of a glass top and bottom, which provides the casting surfaces which are clamped to a flexible gasket that serves as the sides. After a part is made and removed from the mold, the glass molds can be cleaned and reused.

## 13.6 Measuring Techniques

Because of the criticalness of the tolerances of a lens, measurement is very critical. There are several techniques that can be used to measure focal length and radius of curvature.

To develop a radius measurement, the simplest way is with a device called a spherometer. This device has one measuring anvil in the center and three others arranged in a circle out from the center. By measuring the distance to the center and three points around the periphery of the lens the radius of curvature can be calculated. Knowing the radius, lens thickness and index of refraction of the lens, the focal length can be inferred.

A lensometer, which is the instrument an optometrist uses, through the use of a lens system can determine the focal length and radius of curvature of a lens.

A lens or optical bench again uses a lens system and will produce a very accurate measure of focal length.

A very accurate way of measuring a radius of curvature is by using a master test lens of a known radius the same as the radius as the lens you are trying to duplicate. By bringing the two into contact under light, bands or lines which are interference fringes will be observed. These bands are known as "Newton's rings." Each ring represents the departure of the specimen by one-half of a wavelength of light. If you observe 10 bands from the center spot to the outer edge that means that the space between the master and the lens being tested is equal to 5 wavelengths or about 0.0002 inches (0.005 mm).

# 14

# Joining Techniques

One of the advantages of designing plastic components is that you can often combine two or more functional parts into one piece, saving on manufacturing costs and assembly costs. That being said, you often will have to join the part you are making with others of the same or different materials in order to make the finished product. There are many different techniques that can be used, depending on the materials involved and the functional requirements of the assembly. Some techniques are similar to those used to join metals or other materials, but they will have to allow for some factors unique to the plastic materials. Others will be unique to plastics materials and have their own design guidelines. Some plastics can be joined by a wide variety of techniques and some will work only will a select few. The discussion below is intended to give the reader an overview of the common joining methods available for use along with some suggested design guidelines and a discussion of what materials will work well with each. Lastly, considerations to be taken in design for assembly will be discussed.

## 14.1 Threads and Threading

Mechanical means of assembly generally will work with almost any material and is often the best choice when trying to join dissimilar materials together. We have already discussed one of the most common methods of mechanical assembly, snap fits, in Chapter 9. Another very common mechanical method of joining is the use of threads. In this section we will discuss making threads an integral part of the part being designed. We can also incorporate threads into our design through the use of self-tapping screws and also by molding-in or welding metal threaded inserts into the part, which we will discuss later.

The ease of processing plastics materials makes it possible to incorporate any thread design into a plastic part. In principal, threads incorporated as part of a plastic design work similarly to threads in any other material. In practice, however, the unique characteristics of plastic material behavior will affect how the threads need to be designed. The natural lubricity of many plastic materials is a plus for threaded applications, but shrinkage and creep have to be accounted for.

When designing a plastic thread, traditional forms can be used but the thread performance can often be enhanced by modifying standard forms to account for plastic behavior. Coarse threads are generally preferred due to their higher strength and torque limits. For applications requiring high pullout or that will carry high pressure loads, Acme and buttress thread forms are good selections.

When designing a molded plastic thread, you need to remember that the plastic material will shrink after molding, affecting the thread length, diameter and form. This can usually be compensated for in the tool design. A sharp, pointed thread will be easily bent or chipped. For this reason and the fact that it is very difficult to mold a thread with a sharp point, most molded threads are modified to incorporate a small radius at the tip. If a sharp point is required, machining of the thread is always an option.

Threads are classified as male (threads are external to the part) or female (threads are internal to the part). Both types should be designed with a lead-in thread relief to prevent high stresses at the end of the thread, as shown in Figure 14.1. A minimum of 0.030″ (0.76 mm) should be provided. Also, good design practice warrants including a minimum of 0.005″–0.010″ (0.13–0.25 mm) radius at the root of the threads to reduce stress concentration.

External threads are normally located on the mold parting line or made by a side pull. When the threads are formed this way there is

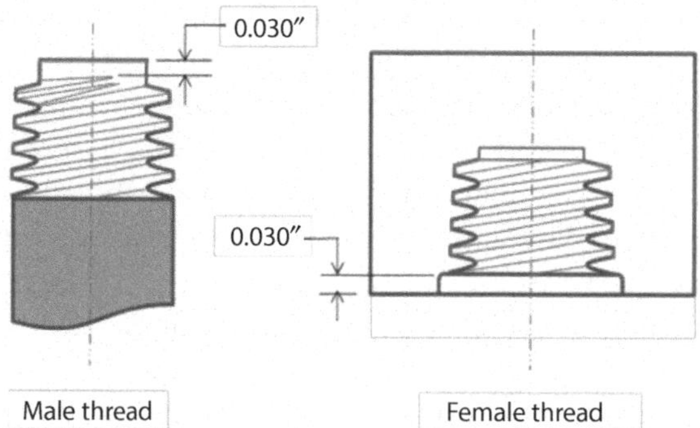


**Figure 14.1** Thread relief.

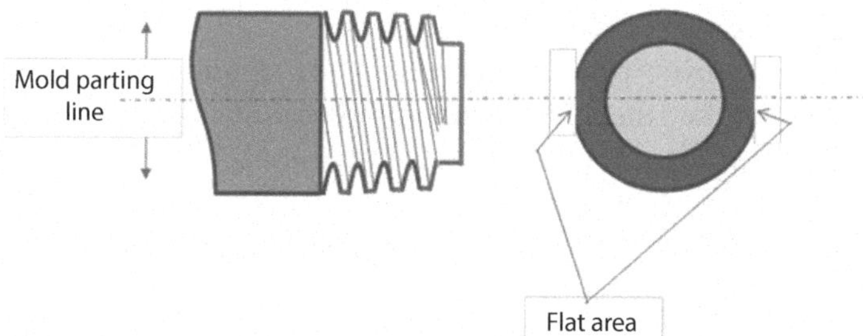


**Figure 14.2** Parting line flats.

always the possibility of parting line flash or mold mismatch causing an area of interference on the threads. This can be alleviated, as illustrated in Figure 14.2, by incorporating a small flat at the parting line which will keep any flash below the thread OD. Additionally, this will make routine mold maintenance much easier. Another option for tool design would be an unscrewing mold. This eliminates the parting line concern, but will have a higher tool and operating cost.

Internal threads are usually formed by either an unscrewing mold or by a collapsible core. Non-sealing threads on thin wall parts in ductile materials can sometimes be directly stripped off of the thread core by the mold ejector system.

With internal threads, anisotropic shrinkage can be a serious concern, particularly on larger parts with high shrinkage materials or glass-filled materials. As the thread diameter becomes slightly oval, it will cause an increase in assembly torque with its mating component and in worse case will not assemble. Uniform mold cooling and gate location are important. Adding glass beads to the material can help produce more uniform shrinkage. For example, if using 10% glass fiber-filled polypropylene, adding an additional 10% glass beads will significantly improve the anisotropic shrinkage. As a last resort, thread blanks can be molded and the thread form post-machined as a secondary operation. There can be a similar situation with male threads, but because the thread inserts don't rotate, they can be modified to compensate for differential shrinkage.

When designing a tapered pipe thread, particular care needs to be taken on the internal threaded part. The taper of a pipe thread acts as a wedge which can generate very high hoop stress. This can result in the cracking of the internal part, particularly at the weld line. To avoid this, a positive stop or shoulder must be incorporated at the bottom of the internal thread, which will prevent the male member from driving too far into it, causing breakage. This is illustrated in the photo of parts with female and male pipe threads shown in Figure 14.3. An approximate evaluation for the hoop stress produced in a plastic boss with internal pipe threads is shown in Equation 14.1 below [65], where $\sigma$ is the hoop stress, $T$ is the assembly torque, $t$ is the wall thickness, $d$ is the diameter and $L$ is the effective thread engagement. The coefficient of friction is assumed to be 0.15.

$$\sigma = \frac{3T}{tdL} \tag{14.1}$$

**Figure 14.3** Parts with female and male pipe threads.

**Figure 14.4** Modified plastic valve body and nut.

Plastic female pipe threads should not be used to join with male metal pipe threads. A male plastic thread can be used with a female metal thread. In all cases, care needs to be taken during assembly not to overtighten the threads.

Thread forms with thicker roots, such as an Acme thread form, are commonly used for plastic load bearing threads in applications such as appliances, valves and pumps. Adding a radius of at least 0.15″ (0.38 mm) to the base of the thread is critical to minimize the stress at the base of the thread during loading. Typically, the standard Acme thread form will be modified for molded plastic to compensate for shrinkage variations from thread to thread and in the case of multi-cavity molds from cavity to cavity. The most common modifications are reducing the thread thickness slightly to give more flank clearance and using a stub design (shortening the thread height) to give more radial clearance as the threads are assembled. Figure 14.4 shows a plastic valve body and nut showing both of these modifications. As you can see, the tops of the threads are flattened to a stub form and the threads themselves are thinner than the spaces between the threads. This will allow very low assembly torque up until the point that the threads tighten.

## 14.2 Self-Tapping Screws

Self-tapping screws are a good choice for single-use or limited-use assembly operations. The advantage of a self-tapping screw is that it eliminates the need to mold a female thread into the part and just requires a hole to accept the screw. This simplifies the tool and tool cost or eliminates the need to tap a thread into the part.

There are two basic types of self-tapping screws that are used in plastic parts, thread-cutting screws and thread-forming screws.

Thread-cutting screws are generally used on rigid, brittle plastics such as highly filled thermoplastics and thermoset materials. Thread-cutting screws are usually designed with a slotted shank and will actually remove material as they cut the threads when they are driven into the pilot hole in the part. A chip reservoir needs to be added at the bottom of the pilot hole to collect the shavings. Because they actually cut threads upon installation, they should not be used for reinstallation. If a reassembly is required, an appropriate sized machine screw should be substituted. The pilot hole for thread-cutting screws should be large enough so that no more than 75% of full thread engagement is reached.

For most thermoplastic applications thread-forming screws that are specifically designed for plastics are typically used. These screws work by having a lobe-shaped shank that pushes material out of the way as they are driven. After installation, the plastic will cold flow around the shank, tightening the screw in place. These screws can normally be reused up to 6 times effectively. If repeated assembly and disassembly are required, metal inserts should be used. An example of this type of thread-forming screw is the PLASTITE® (Research Engineering and Manufacturing, Inc.) screw. Thread-forming screws can induce high stresses into the plastic as the thread is formed and can cause cracking in brittle materials or in parts with weak weld lines.

Figure 14.5 shows a schematic of a self-tapping screw and the boss it will be driven into. The OD is the outside diameter of the screw threads and "d" is the diameter of the pilot hole. When designing a hole for a self-tapping screw, the thread engagement should be at least 2.5 times the OD. The pilot hole diameter should be about 75% to 85% of the OD for soft plastics and 85%–95% of the OD for hard plastics. The screw manufacturer (or the material manufacturer) you choose will give you a specific recommendation for the material you have chosen. The boss diameter should be at least 2 times the pilot hole diameter, and needs to be much larger for small screw diameters. The cored hole should have 1/4° to 1/2° draft per side. The pilot hole needs to be chamfered to a depth of 0.20″–0.30″ starting at the diameter of the screw to minimize the possibility of the boss cracking. Molded bosses typically have weld lines which must be minimized by good mold design and venting and by processing at optimum conditions.

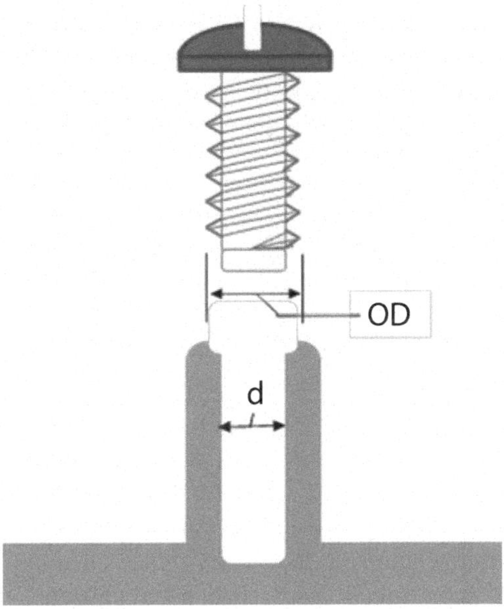


**Figure 14.5** Self-tapping screw.

## 14.3 Metal Inserts

Metal inserts are used when a plastic part is intended to be assembled and disassembled many times or when there are concerns over creep due to the need for a high assembly force. Metal inserts can be installed by incorporating press fits, molding in place, or by ultrasonic insertion.

The press fit is a simple and economical way to attach an insert (or other metal component) to a plastic part but must be done correctly to avoid problems such as the insert loosening over time or the plastic part cracking from the stress caused by the insert. The part must be designed to adequately provide the holding strength that the assembly requires without overstressing the part. It must be manufactured to a tight tolerance to keep the stress within allowable limits. It will change dimension over time due to normal plastic stress relaxation or creep. Creep can cause a press fit that initially provided adequate resistance to torque to loosen to the point that the insert will spin in the part when assembly or disassembly is attempted. This problem can be overcome by incorporating knurls or grooves into the insert. The lower contact area will reduce the assembly stresses and the plastic will cold flow into the spaces in the knurls or grooves over time, locking the insert in place.

When designing the press fit of a metal insert into a boss on a plastic part, the hole size deformation that the operation will create must be calculated to insure that the plastic part grips the insert as tightly as possible without being overstressed. This can be done by using Equation 14.2 below [66]:

$$\delta = b\varepsilon\left(\frac{c^2 - b^2}{c^2 + b^2}\right)\left(\frac{c^2 + b^2}{c^2 - b^2} + \nu\right) \tag{14.2}$$

where $\delta$ is the amount of radial deformation of the boss, $b$ is the radius of the plastic boss (½ of the boss outside diameter), $\varepsilon$ is the allowable strain for the material being used, $c$ is the radius of the insert being used and $\nu$ is Poisson's ratio for the plastic material. Figure 14.6 illustrates an example of how this equation will be used. A part is being designed to be assembled to another part with brass threaded inserts. The plastic material chosen is a 33% glass-filled nylon 6 material. A standard 5/8 inch metal insert has been chosen. The proper hole diameter to provide maximum holding power without overstressing the insert needs to be calculated. Initially, a 1″ diameter has been chosen for the boss to provide a wall diameter less than the thickness of the wall it is joined to, minimizing the possibility of a sink (see Chapter 5 discussion). The allowable strain $\varepsilon$ of 2% and Poisson's ratio $\nu$ of 0.35 are found from the material supplier's data sheets. Substituting these values in Equation 14.3 we find the radial displacement as follows:

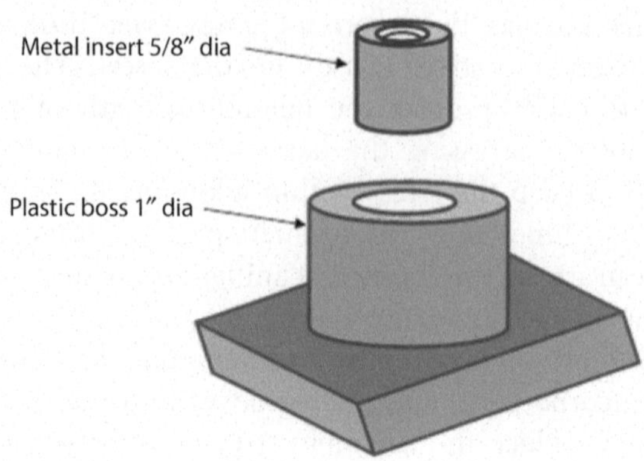


**Figure 14.6** Press fit insert.

$$\delta = (.3125)(.02)\left(\frac{0.5^2 - .3125^2}{0.5^2 + .3125^2}\right)\left(\frac{0.5^2 + .3125^2}{0.5^2 - .3125^2} + 0.35\right) \quad (14.3)$$

$$\delta = .00577'' = \text{radial clearance}$$

The boss hole diameter should be designed to provide an interference fit of 0.011″ with the insert OD ($2\delta$ or twice the radial deformation). If this does not give the design enough holding force, the boss diameter can be increased or the insert length increased. Increasing the plastic boss diameter will increase the amount of interference that is acceptable. In the above example, if the boss was increased to 1–1/4″ the allowable interference would increase to .012″. If there is room in the design envelope, increasing the insert length is a more effective way of accomplishing this. Insert manufacturers can provide estimated pullout forces of various materials with their inserts, but the best data will come from the testing of a prototype.

Molded-in inserts, as the name implies, involve placing the metal insert into the mold and encapsulating the insert with molten plastic in the molding process. This process can be used with both thermoplastic and thermoset materials. An advantage of this method is that you can mold-in inserts with significant knurls or undercuts, greatly enhancing the pullout force of the insert. The process of loading the inserts into the mold will increase cycle time and increased the potential for tool damage, but generally will be more economical than a secondary operation, particularly on high volume parts where it makes sense to automate the insert loading operation. When specifying an insert to be molded in place, care should be taken that the inserts do not have any sharp edges or corners in contact with the plastic that can act as stress risers and cause the part to fail prematurely. Be particularly careful with knurled inserts as they will generally have sharp, rough points unless carefully specified otherwise. Brass is the most common material used, although almost any metal will work. The photo in Figure 14.7 gives an example of molded-in inserts, with the part on the left having a brass insert and the one on the right a stainless steel one.

The third method of installing inserts is ultrasonic insertion, which will be covered in the next section on ultrasonic welding.

**Figure 14.7** Molded-in metal inserts.

## 14.4 Ultrasonic Welding

Ultrasonic welding is a fast, reliable method that can be used to join the same or very similar plastic parts together. It is one of the most widely used joining techniques for thermoplastics and will produce a permanently bonded and aesthetically pleasing joint. The ultrasonic welding process converts electrical energy into mechanical vibrations at ultrasonic frequencies. These vibrations are transmitted through one of the plastic parts, causing friction between two mating parts. This friction causes the plastic to melt. The parts are held in a fixture under pressure while the energy is applied. After sufficient melting occurs, the energy is shut off and pressure is maintained until the melted plastic solidifies, at which point the assembly can be removed. The welds are consistent and strong and a properly designed and processed weld will be as strong as the material itself. Standard ultrasonic welding frequencies are 20 kHz and 40 kHz. The welding cycles are typically very fast, around 2 seconds or less, but can get a little longer on large parts that might require a little more cool time. The process is very clean, but hearing protection should be worn around a welding operation.

Materials must be rigid enough to transmit the ultrasonic energy and they have to be compatible in order to weld together. A material's weldability can be enhanced by adding a small amount of filler to increase the rigidity. When designing an ultrasonic welded joint, your location of the joint must consider that stray energy can damage delicate features in close proximity to the weld area. When welding hygroscopic materials,such as nylon, it is important to weld parts that are dry to prevent weak or brittle welds. This can best be done by welding parts directly from the molding machine.

To optimize the welding process the ultrasonic horn which directs the energy from the welding machine to the point of welding must be correctly designed and sized, taking into consideration the size of the welding machine, the material being used and the size and geometry of the weld. The welding machine supplier is the best resource in this regard.

The most important factor in determining the quality of an ultrasonic weld is the joint design. The two principal joint types are the energy director joints and the shear joint. The design will depend upon the type of material to be welded, the part geometry and the product end use requirements.

An energy director is a raised triangular section that is molded onto one of the two parts to be welded together. The purpose of an energy director is to concentrate the ultrasonic energy and focus it on the area to be welded, causing a rapid melting and welding in the area intended. Energy director joints are normally used when welding amorphous plastics.

Figure 14.8 shows an example of an ultrasonic energy director being used to make a butt weld between two plastic parts. In the circle on the left is a detail of the actual energy director. The triangle typically will be at 90° but may be reduced to 60° if used on a thin-walled part. The height of the triangle should be about 1/8 of the nominal part wall thickness. The base of the triangle, as shown in the center figure, should be about 1/4 of the nominal part wall thickness. As the ultrasonic

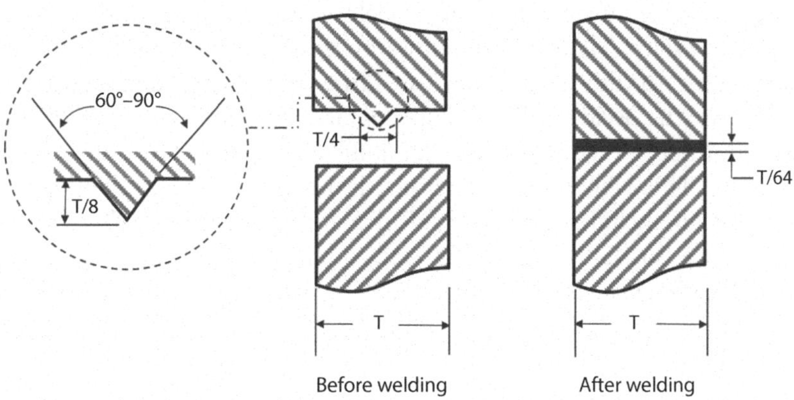


**Figure 14.8** Butt joint with an energy director.

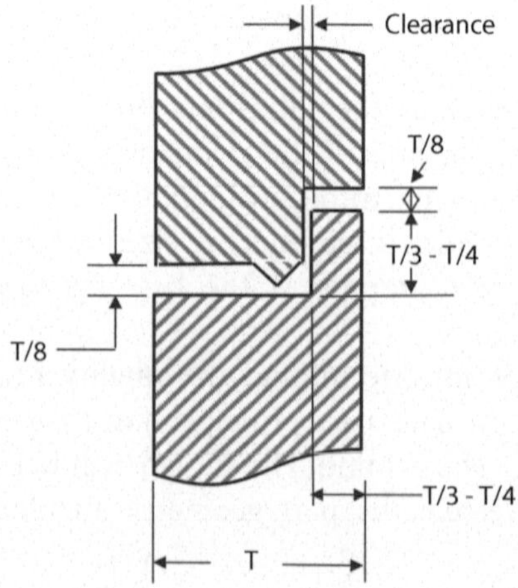


**Figure 14.9** Step joint with an energy director.

energy is applied it will melt the energy director and the material it comes in contact with and gives a uniform weld between the two parts.

Figure 14.9 illustrates a step joint using an energy director. This joint can be used to obtain a leak-tight seal. Notice the clearance between the two parts to the right of the energy director. The clearance will function as a flash trap, providing a space for excess melted material. Clearances are typically 0.002″ to 0.004″ (0.05–0.10 mm).

Shear joints are commonly used to weld together semicrystalline materials such as polyesters and nylons. When using an energy director design with these materials they often will not achieve a strong weld due to their sharp melting points. The melted material flowing from the energy director quickly resolidifies before melting and welding to the opposite surface. Figure 14.10 shows an example of a typical ultrasonic shear joint. In the shear joint, as ultrasonic energy and pressure is applied to the weld area, a small contact area is initially melted and then continues down the rest of the joint area as one part is forced into the other. This will result in a strong, leak-tight seal being developed. A good design includes a small lead-in a minimum of 0.020″ (0.5 mm) deep and a 30–45° angle at the start of the interference area as indicated. Interference should be from 0.006–0.008″ (0.15–0.2 mm) for thin parts (1/2″ or 13 mm or less) up to .016–.020″ (0.4–0.5 mm) for thicker parts. A flash trap can be provided in the joint design similar to the energy director design.

Ultrasonic staking is a good method for permanently joining parts of dissimilar materials together. Figure 14.11 illustrates this concept.

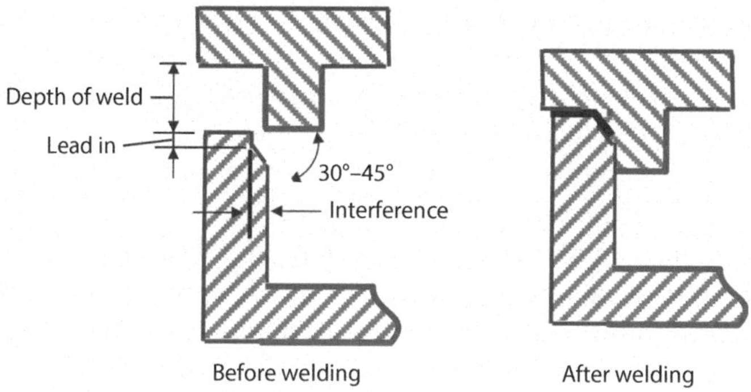


**Figure 14.10** Ultrasonic welding shear joint.

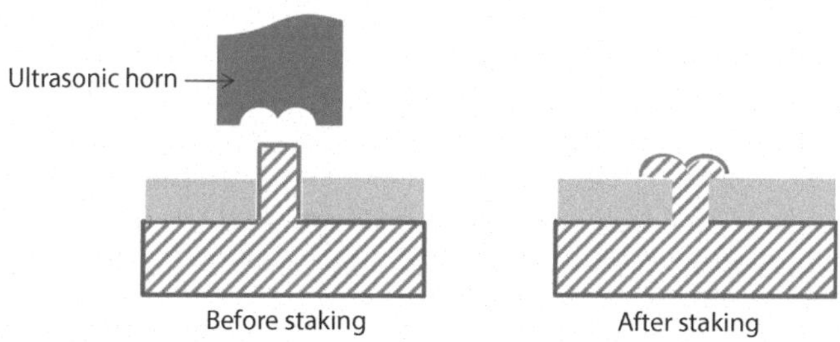


**Figure 14.11** Ultrasonic staking.

A (usually) round post is molded onto one of the mating parts. The other part is assembled over the post and a specially designed ultrasonic horn applies energy to the post, swaging it over the top of the second part, locking it in place. This assembly cannot be taken apart without breaking the post.

Ultrasonically assembling a threaded insert into a part is similar to staking, except that instead of a post, a hole is molded or drilled into the part to accept the insert and an ultrasonic horn applies ultrasonic energy and pressure to the insert. The insert transmits this to the plastic, melting the plastic around it as it enters the hole. This technique will result in lower stress than a cold staked insert, and because the plastic melts and flow around the insert and then resolidifies, the insert is firmly locked in place.

## 14.5 Vibration and Hot Plate Welding

Vibration welding is a good technique for joining together large thermoplastic parts, particularly those that are too large for ultrasonic welding. Figure 14.12 shows a schematic of a vibration welding machine. In vibrational welding, one component is held fixed in the stationary fixture while the second component is rapidly vibrated back and forth while pressure is applied, pushing the parts together. The vibrations produce friction between the plastic parts which cause them to melt and form the joint. The vibration frequency produced by the electromagnet is either 120 Hz or 240 Hz (in the US). An alternative design uses a hydraulic force to generate the vibration frequency. This design makes it easier to obtain higher frequencies [67]. Amplitude of vibration is typically between 1/8 to 1/2″ (3.2–12.7 mm).

Joint designs are typically L-shaped or T-shaped. Figure 14.13 shows a schematic of a typical L-joint. In the two figures on the left, you see an L-shaped joint. As the weld is made, some plastic will be displaced out the ends of the weld, leaving a bead of weld flash. If this is a cosmetic or tolerance issue, the flash would have to be trimmed off. An alternate design is shown in the two figures on the right. Flash traps have been added to the design to capture the flash and keep it internal to the part. These may have to be modified once the actual amount of flash the process will generate is determined. Also, to achieve good and consistent weld quality, the assembly fixtures must be designed to clamp the parts firmly as near to the weld area as possible.

Two or more separate weld areas can be incorporated on parallel planes to the direction of motion in the same assembly, allowing the simultaneous production of multiple sealed areas in the part.

Hot plate welding, also called butt fusion, is an economical joining method for permanently joining two parts together along a single plane.

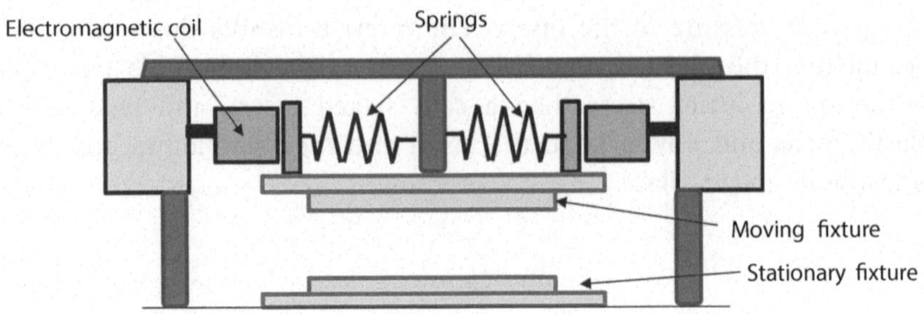


**Figure 14.12** Schematic of a vibrational welder.

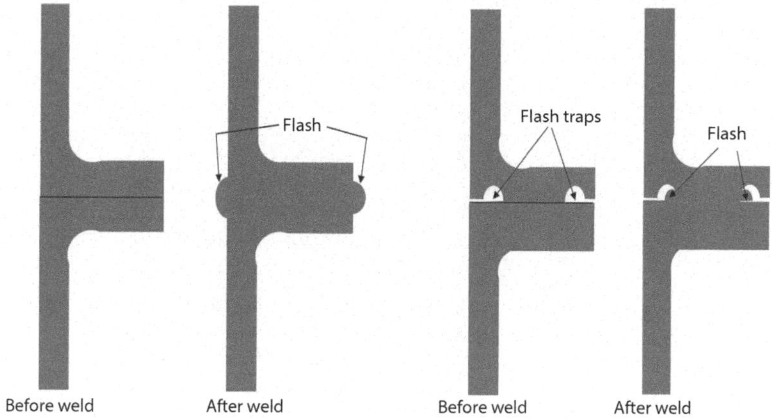


**Figure 14.13** L-shaped joint design with and without flash trap.

It is a very common method used to join plastic pipe. In the basic hot plate welding process, a flat metal plate is heated and brought into direct contact with the two plastic components to be assembled, causing melting on the surfaces to be assembled. The metal plate is removed and the parts are pushed together and held under low pressure. The melt from both sides will join and as the plastic cools, a weld will form. The quality of the weld is dependent on the cleanliness of the plastic part and the welding plate, the welding temperature and time, and holding the parts stationary during the cooling process. In a properly made weld, the strength of the joint will equal that of the material it is made of. There normally will be a welding bead or flash around the welding area. The photo in Figure 14.14 illustrates a finished butt weld that is cooling in the welding fixture. The welding beads formed around the weld area are clearly visible.

This method can be applied to other simple geometries besides a flat plane. An example of this is the socket fusion of pipe fittings where the socket of the pipe fitting is melted with a round heating tool that is inserted into the fitting socket.

An interesting modification to this process is infrared radiant butt fusion. In this process an infrared heater is brought into close proximity to, but not contacting, the plastic parts to be joined. The infrared heater softens the plastic, which is then pushed together and held until the weld is cooled. The advantage of this process is that nothing directly contacts the plastic, making for a cleaner weld, and it tends to be more repeatable and produces a smaller weld bead. The disadvantages are the

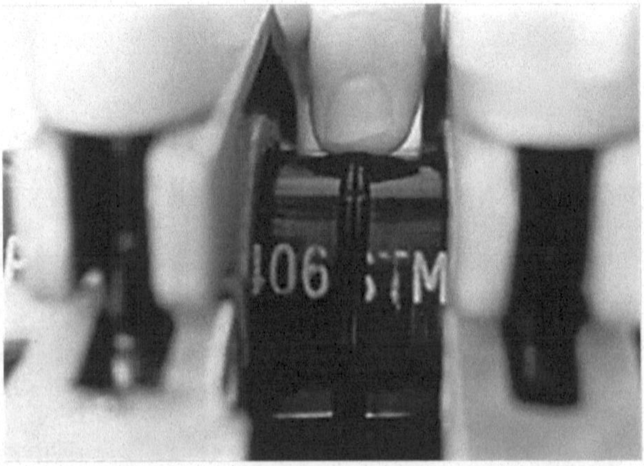


**Figure 14.14** Finished butt weld joint. ©2003 Plastics Pipe and Fittings Association

equipment is more expensive and the process is more complex [68]. It is commonly used in ultrahigh purity fluid systems in polypropylene and PVDF materials.

## 14.6 Spin Welding

Spin welding is a fast and cost effective technique for permanently joining together circular parts or surfaces. It can be used on most thermoplastics, but rigid materials work best. Frictional heat is generated by rotating one of the parts at high speed and pressing it with controlled pressure into a second stationary part. The frictional heat will melt both part surfaces. Once a melt layer is established, the rotation is stopped and the plastic resolidifies, welding the two parts together. Strong hermetic welds can be achieved, but precise location of the two parts diametrically relative to each other is difficult. Typical weld times will be a few seconds.

For prototypes or low volumes, equipment as simple as a modified drill press and fixture can be used. For higher volume requirements, custom spin welding equipment can be built incorporating automation and process control.

The welding joint can be flat or angled. Parts designed for spin welding often have alignment features such as a tongue and groove to provide a guide and uniform bearing surface. As with some of the other methods discussed, flash traps can be incorporated into the design to

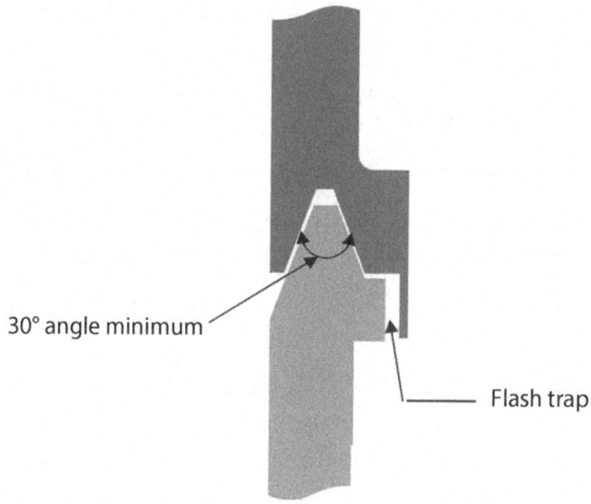


**Figure 14.15** Tongue-and-groove joint.

avoid visible welding flash. The incorporation of a flash trap in a tongue-and-groove joint design is illustrated in Figure 14.15. When designing a spin weld joint, the total surface area of the weld joint should be at least 2–1/2 to 3 times the wall thickness to achieve optimum joint strength.

## 14.7 Solvent and Adhesive Bonding

Adhesive and solvent bonding are among the most versatile methods of making a permanent assembly. Adhesives are one- or two-part glue systems that adhere to the material surfaces and cure to form a bond between the materials that they are attached to. Solvent bonding joins one part to another by softening the parts with a volatile solvent and the parts stick or bond together when the solvent is flashed off and the parts harden.

Adhesives allow great flexibility in design because they can be effectively used to bond a plastic material to itself, to other plastics or to an array of other substrates, including metals, wood, glass and ceramics. Adhesives must be used with care as they usually stick well to anything, intended or not and some of them can be hazardous. A Material Safety Data Sheet should be obtained from the manufacturer or supplier prior to use for safe handling instructions.

Adhesives that can withstand one or more of the following conditions, mechanical loading, elevated temperatures, and creep, are generally classified as structural adhesives. Below is a discussion of some of the most common ones.

Polyurethanes make a tough, high strength joint with good impact resistance. They exhibit low temperature flexibility but have limited moisture resistance. They are normally a two-part system, an isocyanate and a multifunctional alcohol. The crosslinking reaction starts as soon as the components are mixed and the mixed adhesive has a relatively short pot life. The joint normally has to be fixtured to hold in place until it is cured.

Epoxy adhesives possess high strength and high temperature resistance and are widely used in structural composites. They have poor impact resistance and tend to be brittle. Epoxies are a two-part system, the epoxy and a hardener which opens up the epoxy rings. The adhesive can be cured at room temperature or at elevated temperatures. Again, the joint normally has to be fixtured to hold in place until it is cured. Epoxies normally are relatively inexpensive.

Cyanoacrylates (Super Glue®, Krazy Glue®) possess very high adhesive strength but have limited service temperature (about 200 °F or 93 °C), low impact strength, and low resistance to moisture. They are a one-part system that cures in the presence of very small amounts of moisture as the amount typically present in the air. Cyanoacrylates exhibit a very rapid cure time which makes fixturing the assembly usually unnecessary. They tend to be relatively expensive.

Phenolics and urea formaldehydes are thermosetting resins available in one- and two-component systems. They cure with heat and are strong but brittle. They are commonly used in chipboard and plywood and are relatively inexpensive.

Silicones have very high heat resistance, good impact resistance, good low temperature flexibility and good sealing capability against a wide range of substrates. They are also resistant to water and most solvents. They are available in one-component (moisture cure) or two-component systems. They can be formulated to cure at room temperatures (RTV) or at elevated temperatures. Silicones have very long cure times and will normally need to be fixtured to hold the joint stationary until cured. They are relatively expensive.

When selecting an adhesive, the material selected needs to be compatible with both the materials being selected and the requirements of the application. The chemical constituents of the adhesive and the cure temperature cannot exceed the capabilities of the plastic substrate and the cured adhesive needs to function in the mechanical, temperature,

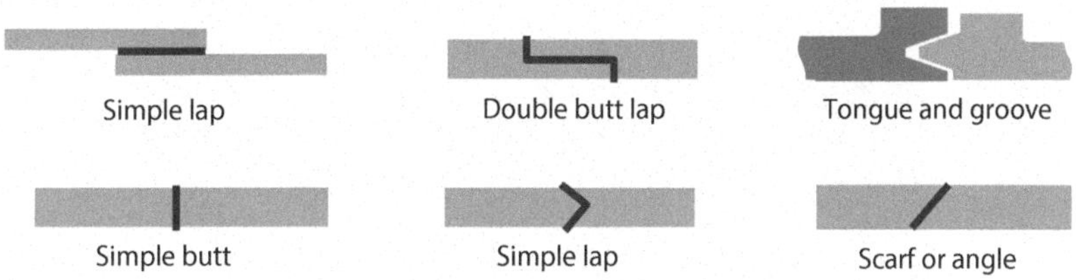


**Figure 14.16** Common adhesive joint designs.

chemical and environmental operating conditions of the finished assembly.

The joint design needs to provide adequate surface area to provide the strength required. Adhesive suppliers can provide data on relative joint strengths of their materials. The joint surface should not be smooth (either a rough mold surface or a secondary operation) and needs to be clean and free of any contaminates when the adhesive is applied. Figure 14.16 illustrates a number of common adhesive joint configurations.

Another method of adhesive bonding is with the use of hot-melt adhesives. Hot-melt adhesives are thermoplastic materials that are heated and softened in a heated dispensing tool which melts the resin and forces it out the tip or a hot air welder that melts the plastic by forcing hot air over it. Common hot-melt adhesives include EVA (ethylene vinyl acetate), PE (polyethylene), nylons, and amorphous polypropylenes. Common applications include sealing packages and attaching light objects to another substrate.

The hot air welder is also commonly used in piping applications for low pressure joints and repair in PVC, PP and PVDF material applications. An example of this is shown in Figure 14.17.

For materials susceptible to oxidation, the welding can be done using a heated inert gas like nitrogen.

Solvent bonding is a very inexpensive way to join plastic parts together. Amorphous thermoplastics are more suitable for solvent bonding than crystalline materials. Materials with excellent resistance to solvents, such as polyethylenes, cannot be bonded by this method.

In solvent bonding, solvent is applied to the joint area of one or both of the components to be assembled, softening the plastic. The two parts are held together under pressure and as the solvent evaporates, the softened plastic hardens into a permanent bond. The pressure and time required will depend on the plastic and solvent used and the joint design. Solvent bonds work by the solvent actually dissolving the plastic

**Figure 14.17** Hot air welding of PVC pipe. ©2003 Plastic Pipe and Fittings Association

it comes into contact with, so care must be taken as to the amount of solvent applied and the length of time it remains in contact with the parts. Solvent bonds will usually harden quickly but require 24 hours to develop optimum strength. A well-made solvent joint will have strength as strong as the material it is made from.

Solvents should be used only in well-ventilated areas and the MSDS safety data sheet for the solvent should be consulted prior to use.

## 14.8 Bolt and Screw Assembly

Standard metal fasteners provide a simple and convenient way to assemble parts together. Plastics can be joined to plastics or other substrates by this method. This technique lends itself well to multiple assemblies and disassemblies, and standard bolts and screws can be used. When designing for this assembly method, care has to be taken to ensure that the plastic parts being assembled are not subject to excessive forces such as compressive forces due to assembly torque. Figure 14.18 shows an example of poor and good designs for reducing compressive forces on a bolted together assembly. In the figure at the left, the small washer being used results in the tightening force being concentrated over a small area of the plastic part it is being tightened against. The figure on the right uses a much larger washer with four times the surface area, which will distribute the force over a much larger area. When specifying this type of assembly it is good practice to specify the amount of assembly torque to be used and the ideal size of washer for the plastic to be used can be easily calculated.

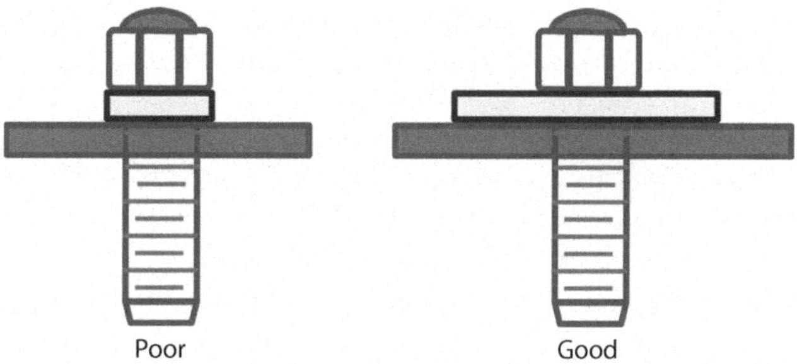


**Figure 14.18** Distribution of assembly torque.

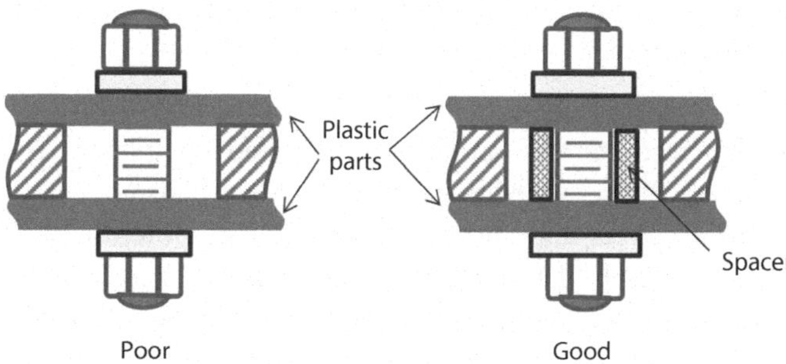


**Figure 14.19** Design to minimize bending stress.

When assembling two plastic parts together with nuts and bolts, situations that subject the plastic parts to bending or tensile forces should be avoided. In the figure on the left in Figure 14.19, as the bolts are tightened the plastic parts will see a bending stress. In the figure at the right, a spacer has been installed to support the plastic parts and the torque applied by the bolts will be opposed by compressive forces in the plastic.

When torque is applied as parts are assembled, it will result in a compressive stress on the part. The part will have an initial deformation but over time the plastic will continue to deform (creep). This deformation will be greater at higher torques and higher temperatures. This can be minimized by distributing the load over a wider area, as shown previously. This will result in a lowering of the torque. In applications where the amount of assembly torque is critical, this can be an issue. By incorporating a spring washer as shown in Figure 14.20, as the plastic part deforms, the spring will compensate and keep the screw or bolt tight.

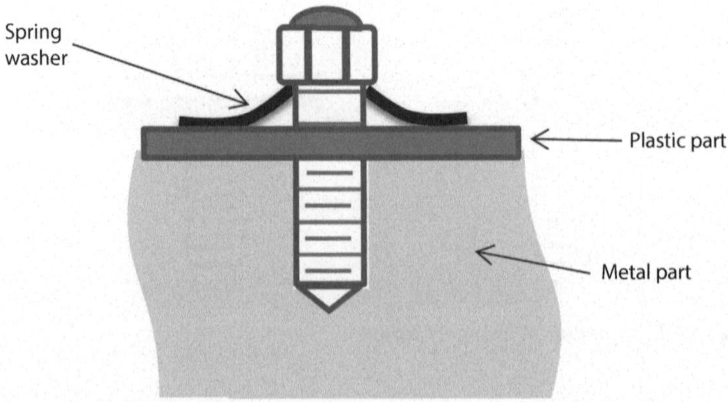


**Figure 14.20** Design to minimize creep effects.

Use of flat-head screws should be avoided in plastic parts. The tapered head will cause a wedging action when the fastener is tightened, which will result in a high hoop stress and can potentially cause the part to fail, particularly if there is a weld line in the hole location. If the fastener needs to be flush with or below the surface, it is better to countersink a hole to accommodate the head of the fastener, being sure to include a radius at the bottom of the countersink.

# 15

# Product Design Process

Plastic product design technology has advanced over the last twenty years as much if not more than plastics materials technology and process technology. Advances in computer-aided design technology, finite element analysis and process analysis and prototyping technology has made the design process faster and more precise, but to take full advantage of this technology, the basic requirements of a good design process are still critical to a successful design.

## 15.1 Design Process

A good design process will involve a team effort and a well-thought-out design strategy to produce optimum results. When designing plastic parts the team should include members of all of the part stakeholders, including design, manufacturing, distribution, marketing and sales. Depending on the part the team could include designers, stylists, design engineers, product managers, manufacturing engineers, tool and die makers, material suppliers, logistics specialists, and last, but definitely not least, customer representatives. Manufacturing engineering includes tooling engineering, process engineering and quality engineering. In a larger company and a large project, this list could grow, in a small company and/or small project one person may of necessity wear several hats, but the important thing is that all interests are represented. By far, the easiest and least costly time to modify a part design is as you are doing it rather than after the first part has been made.

What initiates the need for a product design? It could be driven by a company's strategic initiative, a product line extension, a new innovation, or a request from the customer. What these all have in common is that they all come from someone recognizing a need for the product. This is the action that starts the process in motion. The product champion, who could be the product manager responsible for the strategic initiative, the engineer who developed a new innovation, a sales manager championing a customer request, or anyone with a vested interest in the product, must develop a project proposal. This is the point that the product design team starts to be formed. To develop the project proposal, the product champion will have to get the necessary inputs to put together a project proposal. The project proposal usually will include a product concept, an estimate of the cost to make the product and the price it can be sold for, an estimate of the resources that will be required and a description of how the product will fit within the company's overall marketing strategy. This information will permit a preliminary financial analysis.

The one input that is essential in all cases is to obtain the voice of the customer. This early input will maximize the opportunity for the product to capture the customer's preferences and expectations and to be accepted by the marketplace as it is initially designed and built. This input can be obtained in many ways such as focus groups, surveys, visits with selected customers or having one or two key customers be part of the design team.

The project proposal is the first step of a five-part product development process which is shown in Figure 15.1. The five steps are the development of the product proposal, the definition of the product requirements, the product and process design and development, the process and product demonstration and, finally, delivering a product that delights the customer. The product designer is heavily involved in the first four steps and that is what we will primarily focus on here. You will note the double-headed arrow at the top of each set of activities. Those indicate that in a good process, many of these activities should happen concurrently and feedback from one set of activities will affect the activities of another set. For example, in the project proposal

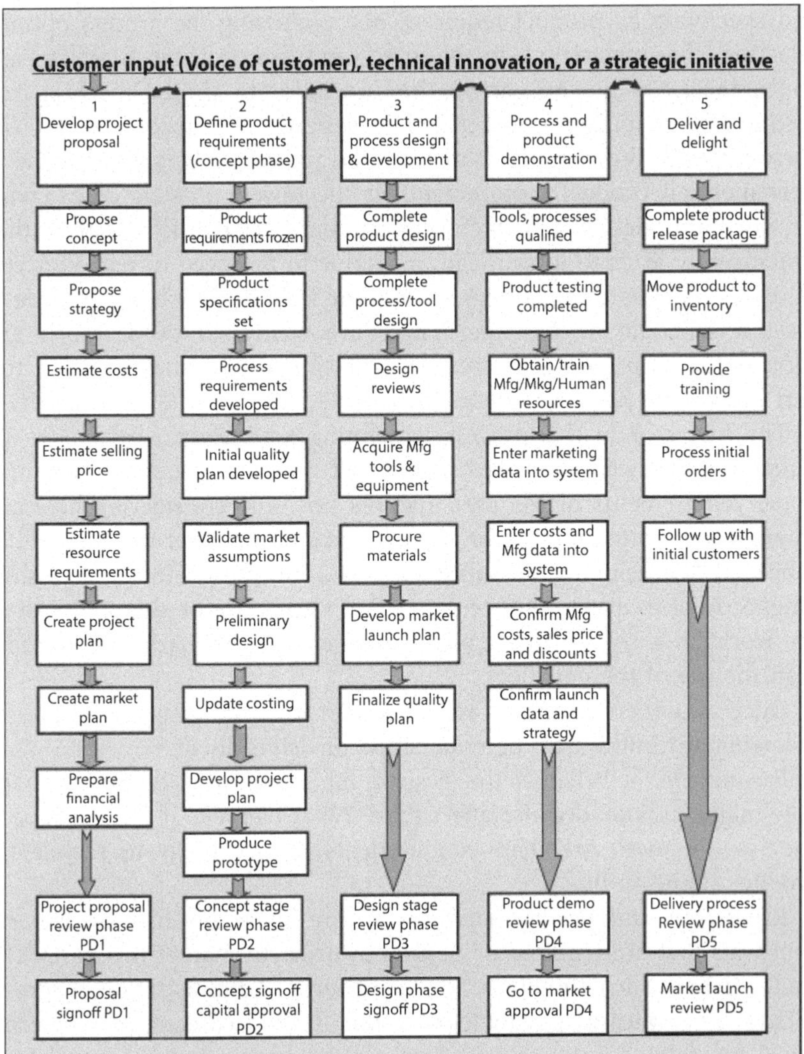


Figure 15.1 Product development process.

phase input from a manufacturing engineer on the costs that would result from a specific visual attribute, such as a texture, could lead the stylist to consider a lower cost alternative.

Once the proposal is approved, the next and probably most important phase of the design is step 2 of the product design process, the design concept phase where all of the product specifications and process requirements are set. Here focus needs to be on completely defining and specifying all product requirements, evaluating the process options, selecting the material(s) to be used and maximizing function and appearance. An integral part of this process is to choose the best path to optimize manufacturing costs. The designer will need to work concurrently with the product champion and manufacturing engineering to fully meet all product requirements at the lowest possible cost. Trying to work the design/manufacturing interface sequentially rather than concurrently leads to longer lead times due to the need to redo concepts or a less than optimum design, as one or both parties have to settle to meet a schedule or timeline. There are a myriad of decisions and choices that must be made, even on a simple part, before the actual part design can be started.

The first step in this process is defining precisely how the part will function (what will it do) and how it will be used. Evaluating the functional requirements of the part involves not only considering how and where it will be used, but also any other conditions that the part will be subjected to, including manufacturing and assembly, storage and shipping. Storing in a non-climate controlled warehouse or shipping around the world in a sea container might expose a part to harsher conditions than the use of the part does.

Once an understanding of what the part must do is agreed upon, the following attributes must be understood or determined:

*Product life* – What is the desired life expectancy of the part and how might its intended use affect that? What happens if the life expectancy is not met? Are there requirements or a strategy to recycle the product at end of life?

*Regulatory and agency approvals* – Are there product or material approvals that are required or desired in order to market this part? This could include such things as Food and Drug Administration Approval (FDA) for applications with food or bodily fluid contact, NSF International for potable water and food processing applications, National Electrical Manufacturers Association (NEMA) standards for electrical equipment and enclosures, and Underwriters Laboratory (UL) for electrical devices and also flammability requirements of a wide range of products. Here you need to be very clear in where the part is intended to be marketed or possibly could be marketed in the future. Many approvals and requirements are very country specific. For example, a part that is to be sold in Canada might require approval from CSA International or in Japan meet the Japanese Industrial Standards (JIS). Approval for many of these standards can require extensive testing and/or approval of manufacturing facilities and quality plans and can take a considerable amount of time. This needs to be factored into the part project plan timeline.

*Safety factors* – What safety factors will you apply to various aspects of the design? Some of these will be driven by code requirements, some by company standards, and some by a risk assessment considering how long the product needs to last and what could happen if it fails.

*Product volumes* – These anticipated product volumes will affect many of the part decisions that have to be made. They are really important. The anticipated volumes will affect decisions made on the process, tool size and design, secondary operations and material choice.

*Cost target* – Knowing approximately where the costs have to be in order to sell the anticipated volume will affect all the same decisions that were affected by product volume. You need to consider the total system cost. If several parts are going into the assembly, the cost targets of all the parts need to be simultaneously and interactively managed. Sometimes, on an innovative product, breakthroughs in manufacturing or material technology will need to be made. This needs to be factored into the project plan and managed and monitored closely.

Next, the actual physical requirements of the part need to be detailed:

*Temperature requirements* – The full range of temperatures that the part will see in operation needs to be considered. Will the part be used only in controlled conditions, will it be used indoors or outdoors or both? Will there be additional temperature exposures due to manufacturing, storage or shipping? If the part is to be distributed globally, what might the range of exposure temperatures be? This is important because many key properties of plastics, such as coefficient of expansion, tensile strength and impact strength, are temperature dependent.

*Mechanical properties* – What are the requirements for physical strength? This will be more precisely defined in the analysis phase of design we will discuss in a moment, but does the part need to have high tensile strength, rigidity, impact strength or hardness?

*Mechanical loading* – Will the part be under any loading? Even a cosmetic part such as a cover might be loaded from an adjacent part or from its operating environment such as someone bumping into it. Also, it could see a load from how it is handled during shipping or how it is packaged. Will the anticipated loads be continuous or intermittent? Are the loads static, cyclical or vibrational? Anticipate all possible sources of impact loading.

*Environmental exposures*– Will the part be subject to outdoor exposure? This will expose it to the potential for UV degradation and variations in the operating relative humidity, which can affect dimensions and strength. This might lead to specifying a UV stabilized material or adding a paint or coating as a secondary operation for protection. If the part is to be used indoors, high intensity discharge lamps or fluorescent lights can emit radiation that can also cause degradation and loss of color.

*Chemical exposure* – Will the part be subject to any chemicals in use, even occasionally? Think not only of any chemicals being used in the direct operation of the part, but also any chemicals it might be exposed to from its immediate environment. For example, a cover on a computer printer might be exposed to grease used to lubricate the printer gears. Also, chemical exposure in the manufacturing environment must be considered. If silicon mold releases are used, they might affect a subsequent coating process. If the part is to have a machining operation, it might be exposed to cutting oils. After the manufacturing process is defined, it is wise to revisit this.

*Electrical requirements* – Will the part be used as part of an electrical device or will it be used in proximity to other parts that are? Are there any concerns about static charge buildup? Does the part have to provide EMI shielding capability? Are there any mandated electrical requirements from codes and standards?

*Aesthetics* – The requirements for how a part is perceived by tactile senses can significantly affect material choices. If the part must have transparency the choices are generally limited to a small number of amorphous plastics. If color must be maintained over time, materials with good weathering stability or with UV stabilizers and/or antioxidants must be chosen. For skin contact, some materials, particularly cellulosics, are said to have a "warm" feel. A few plastics have a unique smell that some may find objectionable. If a very shiny surface is desired, some manufacturing processes, and by extension materials, can provide a better finish.

Based on the determinations made above you can now start to write a product specification, start your product design and begin to establish process requirements, the rest of step 2 and concurrently step 3 of the product design process. It is important to understand that the design will affect the process selected, but also the process selected will affect the part design, so this will be a concurrent and iterative process.

With the information you now have a layout of the part can now be started. This is now almost universally done using computer-aided design (CAD), either in 2D or 3D. 3D design is preferred because of the suite of analysis tools it can be used with and also because of its ability to make mesh models for 3D printing of prototypes. Using the analysis tools you can establish some material choices that will meet the product specifications. With a rough layout, some material choices, and the part attribute information previously gathered, manufacturing engineering will be able to identify a manufacturing process. For example, if the proposed material is a thermoplastic, if the part is very complex and the volume high, injection molding will be a likely choice. If the volumes are very low, machining from bar or sheet stock 3D printing might be considered. If a thermoset is proposed, for a high volume, complex part, transfer molding would be a likely choice and for lower volumes, possibly compression molding or again machining from plastic stock.

It should be noted that this decision is often greatly influenced by the types of manufacturing capability that a company possesses. If a company has only blow molding capability, it is unlikely that an injection blow molding process would be selected. Sometimes it is necessary to think outside of the box, and in order to get the maximum performance/lowest cost, it might be best to contract manufacture a part with a different process or to consider acquiring that process capability.

Each manufacturing process will have its own set of design guidelines which will affect the actual physical shape of the part being produced. When the process has been selected the designer must apply these guidelines to the layout and ensure that they do not interfere with the design intent, the form, fit and function of the part. Many of these guidelines were discussed in Chapter 4, Process Overviews, Advantages and Constraints, and would include such things as fillets and radii, draft angles, shrinkage and tolerances, dimensions of ribs, bosses and intersections, surface appearance, and geometry.

Working with the manufacturing engineers it is time to look at the design of any related parts and see if there is opportunity to combine parts together, such as joining two parts with a molded-in hinge, or to add features that might aid assembly such as locating features or snap fits. Also, now is the time to explore process options and design for manufacturability. Should a texture be molded into the surface of the part or is it better to paint as a secondary operation? Should molded-in inserts or self-tapping screws be used? An extremely tight tolerance can be molded, but sometimes machining it afterwards is a better option. These decisions usually involve evaluations of cost, capability, time, and product performance. This also is the time that an initial quality plan should be developed which will provide valuable input into these types of decisions.

You now can complete the preliminary product design and finalize the manufacturing plan and quality plan. This is a great place to build a prototype with a 3D printing process. The prototype will serve several very important purposes, including serving as a site model for stakeholders to view, providing the opportunity to check fit with mating parts, and although generally not quite the same as a part made by the production intent process from the production intent material, it can also be used for some preliminary testing. It will make any errors in the drawing file apparent and provide the opportunity to correct them before moving further. If it is found that minor changes are required or desired, they can quickly be made and a new prototype produced. This is the end of step 2 in the design process. This is the point at which significant resources are required to move forward and a design review is held with all stakeholders to ensure the design intent has been met and is appropriate.

One more comment on the positive benefits of prototyping. It is not uncommon for design teams to be working on parts of the same project from multiple sites and sometimes even from multiple countries. As one part or subsystem design evolves and changes, it affects parts and subsystems that touch it. Making a prototype part is an extremely effective way of communicating what the change made is to others.

With step two complete, activity moves to completing step 3 and concurrently working on step 4. Any cleanup required for the product drawing is made and if tooling designs are required they are made now. It is extremely helpful to supply a prototype to the tool designer/tool maker to ensure the product design intentions are clearly understood.

As the tool building process proceeds, periodic design reviews will ensure that any changes to the design that the tool maker believes will improve the tool can be validated against the part design intent and the part design remains consistent with the part to be produced. As any tooling is completed and initial sample parts made, the designer will work with the manufacturing engineer to make sure all key design requirements are met and that any changes that might be required to either the part or the drawing so one completely reflects the other are made.

At this point step 4 is completed and the product will be introduced to the marketplace and the customer. If the designer and his team have done the job right, the customer will be delighted that he has the product he needed that performs all the functions required.

## 15.2 Material Selection

The material selection process can be quite daunting. Your design specification for a material might include dozens of properties that are of interest. For an initial scouting of materials, try to narrow down the list to no more than five or six must haves. This will get a set of candidate materials that you can then evaluate in depth.

Be sure to look at similar products that your company makes and also any similar competitive products. The materials they are produced from were generally chosen for good reason.

There are a number of excellent databases that exist that can help you find and compare plastics materials and their properties. Some are generic and give data on multiple materials and manufacturers. I find these a great place to start a material search. Many of these are free, but some of them are wholly or partially subscription services. Other databases are company specific, and while they will have the most extensive data on their own materials, they will not point you to other alternatives. I find these the most useful after a generic search has narrowed the list of possibilities. Table 15.1 gives a list of some selected data sources. Although the list covers a representative range of materials, it is far from complete and you can find a number of additional sources from many more manufacturers.

As you zero in on your requirements, talk to material suppliers. They are a good source of information on applications for the various materials they make. Be sure to ask for examples of similar applications for any materials they recommend. They will not recommend one of their materials for an application that will not work, but the onus is on the designer to be sure the material chosen is the best of the ones that will work.

## 15.3 Design Services

If you lack the resources to design a product by yourself there are a number of resources and services that can supply help with any and all aspects of the project. We already discussed how material suppliers can help in material choices and they often can supply assistance in design reviews and technical analysis.

There are many services that will make prototypes for you either by 3D printing or almost any other process. Although convenient, if you are continually designing parts, your own printer can be a very good

**Table 15.1** Plastics materials databases.

| Website | Material Types |
|---|---|
| Generic | |
| www.campusplastics.com | all |
| www.ides.com | all |
| www.matweb.com | all |
| www.materialdatacenter.com/pt | all |
| Company Specific | |
| www.ashland.com | thermoset polysters |
| www.axial.com | vinyls |
| www.celanese.com | engineering plastics |
| www.covestro.com | engineering plastics |
| www.dow.com/performance plastics/ | engineering plastics |
| www.dupont.com | engineering plastics |
| www.eastman.com | polyesters,cellulosics |
| www.exxonmobilchemical.com | polyolefins |
| www.fpusa | polyolefins, PVC |
| www.lyondellbasell.com | polyolefins |
| www.oxy.com | vinyls |
| www.plenco.com | phenolics |
| www.plexiglass.com | Acrylics |
| www.rtpcompany.com | filled plastics |
| www.sabic.com | engineering plastics |
| www.shintechnic.com | Vinyls |
| www.trinseo.com | Stryenics, engineering plastics |
| www.novachem.com | polyethylene, styrenics |

investment. If you have a functionally demanding part, you might want to utilize a rapid prototype manufacturer who can make your part in your intended production process.

If you need actual design resources, there are many companies that will contract to do the physical design, computer analysis, make prototypes, and even short run production quantities. Try to find someone who has experience in your industry as they will be more aware of specific requirements related to your particular product.

If you need help in the design process, particularly with very large or very innovative projects, there are companies that will work with you to develop your own process or work with you as they follow their own process such as the Eureka Ranch in Ohio and IDEO headquartered in California.

As you work through your design, utilize all of the resources available to you. Your material supplier, your production equipment supplier, and relevant approval and testing agencies, all generally will be willing to talk with you and offer ideas and suggestions.

# Appendix A

## Thermal Properties of Selected Generic Materials*

* These are average values for molding compounds and meant for comparative purposes only. For actual design calculations and intent, consult material suppliers data sheets for properties of specific grades of resins.

| | Specific Gravity | CLTE (in/in/°F | HDT @264 psi (°F) | HDT @64 psi (°F) | Melting Point (°F) | Processing temp (°F) | Tg (°F) | UL 94 Flammability** |
|---|---|---|---|---|---|---|---|---|
| Acetal homopolymer | 1.42 | 5.6–6.1 | 212 | 320 | 352 | 410–446 | | HB |
| Acrylic (PMMA) | 1.19 | 7.5 | 195 | 199 | 400 | 440 | | HB |
| HDPE | 0.96 | | | 158 | 263 | 450–475 | | |
| Nylon 6/6 | 1.14 | 5.6–6.1 | 158 | 392 | 505 | 535–590 | 140 | HB, V2 |
| Nylon 6/6 33% glass | 1.39 | 1.0–4.6 | 486 | 502 | 504 | 540–590 | 176 | HB |
| PBT | 1.31 | 5.1–7.2 | 122–131 | 239–302 | 436 | 455–500 | | HB |
| PBT 30% Glass | 1.53 | 1.1 | 392 | 428 | 437 | 464–482 | | VO |
| Phenolic | 1.41 | 3.2 | 314 | | | 300–360 | | HB |
| Polylactic Acid (PLA) | 1.28 | | | 112 | | 345–400 | | |
| Polypropylene | .915–.916 | 10 | 208 | 131 | | 446–500 | | |
| Polystyrene | 1.05 | 3.9–5.0 | 162 | 184 | | 390–450 | 212 | |
| Polycarbonate | 1.2 | 3.8–5.0 | 270 | 280 | | 500–600 | 297 | HB |
| Polysulfone | 1.24 | 3.1 | 345 | | | 625–725 | | HB |
| PPS | 1.34 | 2.8–6.4 | 203 | 320 | 536 | 590–608 | 194 | VO |
| Polyimide (TP) | 1.37 | 2.6 | 464 | 505 | 730 | 720–810 | 464 | VO-5VA |
| PVC | 1.42 | 4.5–5.5 | 140 | 156 | 153 | 390–410 | 180 | VO |

**Actual flammability can vary greatly between different grades of the same material.

# Appendix B

## Properties of Selected Structural Components

A = area, c = distance from neutral axis to extreme fiber, Na = neutral axis, I = moment of inertia, and Z = the section modulus.

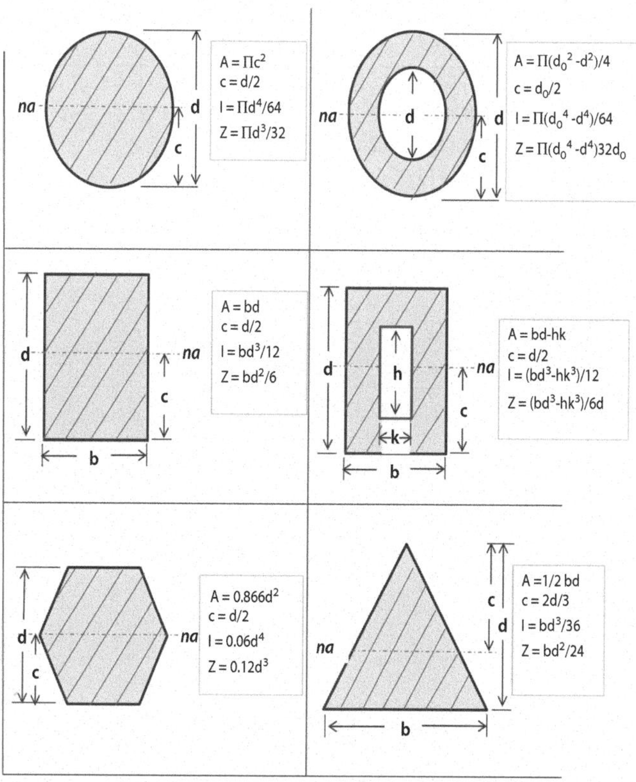

A = area, c = distance from neutral axis to extreme fiber, na = neutral axis, I = moment of inertia, and Z = the section modulus.

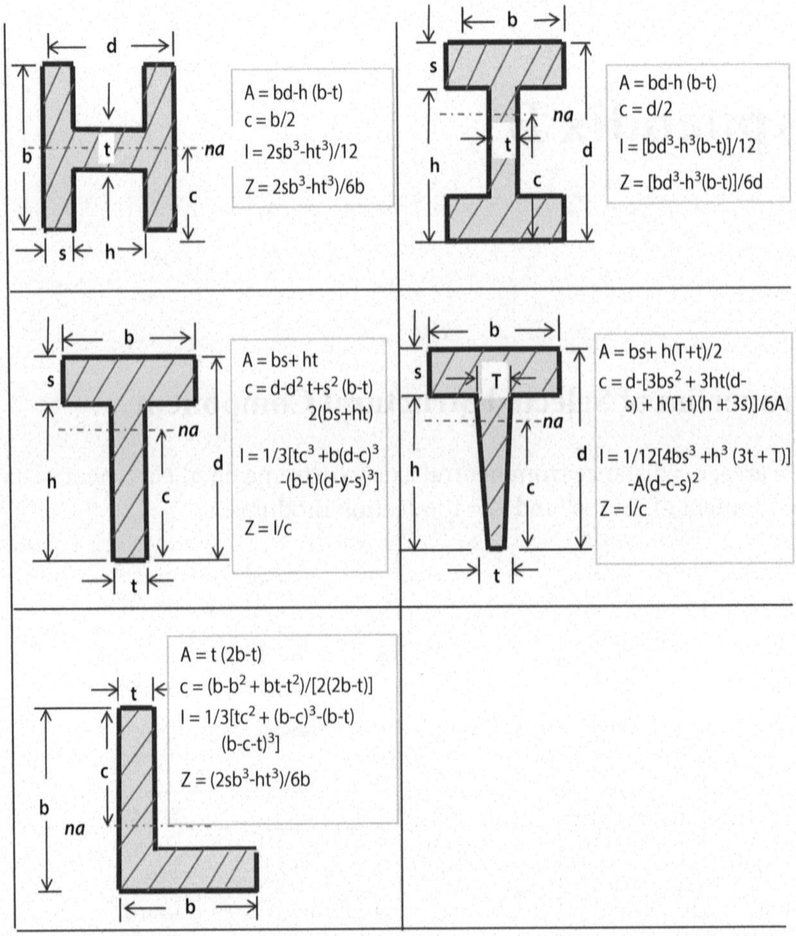

# Appendix C

## Common Abbreviations for Plastic Materials

| | | | |
|---|---|---|---|
| ABS | Acrylonitrile butadiene styrene | PEK | Polyarylether ketone |
| ASA | Acrylonitrile styrene acrylate | PEI | Polyether imide |
| CA | Cellulose acetate | PES | Polyether sulfone |
| CAB | Cellulose acetate butyrate | PF | Phenolic |
| CP | Cellulose propionate | PFA | Perfluoroalkoxy |
| CPVC | Chlorinated polyvinyl chloride | PHA | Polyhydroxyalkanoate |
| DCPD | Polydicyclopentadiene | PI | Polyimide |
| EP | Epoxy | PLA | Polylactic acid |
| FEP | Fluorinated ethylene propylene | PMMA | Polymethyl methacrylate(acrylic) |
| HDPE | High density polyethylene | POM | Polyoxymethylene (acetal) |
| HIPS | High impact Polystyrene | PP | Polypropylene |
| LCP | Liquid crystal polymer | PPE | Polyphenylene ether |
| LDPE | Low density polyethylene | PPO | Polyphenylene oxide |
| LLDPE | Linear low density polyethylene | PPS | Polyphenylene Sulfide |
| MF | Melamine formaldehyde | PS | Polystyrene |
| PA | Polyamide (nylon) | PSU | Polysulfone |
| PAI | Polyamide imide | PTFE | Polytetrafluoroethylene |

| | |
|---|---|
| PB | Polybutylene |
| PBI | Polybenzimidazole |
| PBT | Polybutylene terephthalate |
| PET | Polyethylene terephthalate |
| PETG | Polyethylene terephthalate glycol |
| PC | Polycarbonate |
| PE | Polyethylene |
| PEEK | Poly ether ether ketone |
| PUR | Polyurethane (thermoset) |
| PVC | Polyvinyl chloride |
| PVDF | Polyvinylidene Fluoride |
| SAN | Styrene acrylonitrile |
| TPU | Thermoplastic polyurethane |
| UF | Urea formaldehyde |
| UP | Unsaturated polyester |
| VE | Vinyl ester |

# References


1. HBJ Publications, *50 Years of Progress in Plastics*, 1987 Resin Publications, Inc., page 25.
2. S.L. Simon, *Measurement of the Glass Transition Temperature*, Department of Chemical Engineering, Texas Tech University, Lubbock, TX. ANTECH 2007.
3. D.V. Rosato, D.P. Dimattia, D.V. Rosato, *Designing with Plastics and Composites*, Van Nostrand Reinhold, p 155.
4. A. Brent Strong, *Plastics Materials and Processing* (3rd Edition), Pearson-Prentiss Hall, p121.
5. N. Rao, K. O'Brien, *Design Data for Plastics Engineers,* Hanser/Gardner Publications, p65.
6. N. Rao, K. O'Brien *Design Data for Plastics Engineers,* Hanser/Gardner Publications, p66.
7. A.H. Tullo, Metallocenes rise again, *Chemical and Engineering News*, volume 88, issue 42 October 18, 2010, copyright 2015, American Chemical Society.
8. A. Brent Strong, *Plastics Materials and Processing* (3rd Edition), Pearson-Prentiss Hall, p245.
9. *Tenite*[R] *Cellulosic Plastics*, page 3, Publication PPC-100G Eastman Chemical Company June 2007.
10. P.J. Husperi, R. Layton, and M. Matzner, *Liquid Crystal Polymers, Moderm Plastics Encyclopedia* 1990, McGraw Hill,page 42.
11. *Celazol*[R] *U Series brochure*, PBI PerformanceProducts,Inc, Charlotte, NC.
12. *Ingeo: Ingenious materials from plants not oil, internet brochure,* NatureWorks LLC 2014.
13. IDES Prospector[R]- http://www.ides.com/property_descriptions/ASTMD1693.asp- July30,2014.
14. *Mirel*[tm] *Provisional Data Sheet*, Metabolix, Inc. Lowell, MA, March 2012.
15. *Tellene Brochure* - page 4, Tellene SAS, Drocourt, France, 2011.
16. A. Brent Strong, *Plastics Materials and Processing* (3rd Edition), Pearson-Prentiss Hall, p510.

17. #23 –Pressure Forming, *Plastics Technology, October 2005,* Gardner Business Media, Inc. 2015 Cincinnati, Ohio 45244.
18. The Engineers Edge LLC - http://www.engineersedge.com/plastic_thermoform.htm 2014.
19. J.H. DuBois and W.I. Pribble, *Plastics Mold Engineering*, Van Nostrand Reinhold Company, page 72.
20. *General Design Principles for DuPont Engineering,* EI du Pont de Nemours and Company, 2000, page 10.
21. Konica Minolta Blog, http://sensing.konicaminolta.us/2012/09/measuring-haze-with-a-spectrophotometer/, © 2006-2014 Konica Minolta Sensing Americas, Inc.
22. *NIST Special Publication SP250-70, Specular Gloss,*M.E. Nadal, E.A. Early, E.A. Thompson, National Institute of Standards and Technology, 2006.
23. IDES Prospector$^{R}$- http://www.ides.com/property_descriptions/ASTMD1693.asp- June 14,2014.
24. E.P. Popov, *Mechanics of Materials*, Prentice-Hall, Inc, June 1965, page 435.
25. *Modern Plastics Encyclopedia* 1990, McGraw Hill,pages 714-716.
26. E.P. Popov, *Mechanics of Materials*, Prentice-Hall, Inc, June 1965, pages 37,51.
27. *Modern Plastics Encyclopedia* 1990, McGraw Hill,pages 440.
28. *Modern Plastics Encyclopedia* 1981-82, McGraw Hill,pages 495.
29. *GE Engineering Thermoplastics Design Guide*, General Electric Co., Pittsfield,MA, pages 5-52-5-65.
30. *Modern Plastics Encyclopedia* 1990, McGraw Hill,pages 443.
31. A. Brent Strong, *Plastics Materials and Processing* (3$^{rd}$ Edition), Pearson-Prentiss Hall, p167.
32. A. Brent Strong, *Plastics Materials and Processing* (3$^{rd}$ Edition), Pearson-Prentiss Hall, p168.
33. E.P. Popov, *Mechanics of Materials*, Prentice-Hall, Inc, June 1965, page 435.
34. T. Spahr, presentation *Snap Fits for Assembly and Disassembly*, Ticona, November 1991, page 7.
35. *Snap Fit Joints for Plastics- A Design Guide*, Bayer Material Science LLC, February 2013, page 9.
36. *Snap Fit Joints for Plastics- A Design Guide*, Bayer Material Science LLC, February 2013, page 8.
37. *Design Calculations for Snap Fit Joints for Plastic Parts*, Celanese Ticona, 2009, page 14.
38. *General Design Principals for DuPont Engineering Polymers,* E.I. duPont de Nemours and Co. 2000, Module 1 Chapter 6, page 5.
39. Z. Smith and D. Sheridan, *Gear Molding-Where it is Heading*, March 2006, Plastics Technology.
40. *A Guide to Plastic Gearing*, General Electric- GE Plastics, Pittsfield, MA, 2006 page 3.

41. *A Guide to Plastic Gearing*, General Electric- GE Plastics, Pittsfield, MA, 2006 page 4.
42. *General Design Principles for Dupont Engineering Polymers -Design Guide Module 1*, E.I.du Pont de Nemours and company, 2000, page 54.
43. W. Dornfeld, *Gear Tooth Strength Analysis*, Fairfield University, 2004, page 6.
44. W. Dornfeld, *Gear Tooth Strength Analysis*, Fairfield University, 2004, page 9.
45. *Design of Plastic Gears and General Considerations of Plastic Gearing*, SDP/SI,New Hyde Park, New York,2015, page 6,.
46. *A Guide to Plastic Gearing*, General Electric- GE Plastics, Pittsfield, MA 2006 page 5.
47. *Gear Quality Number (Q Number)* Sandmark Global, Midlothian, VA August 22, 2011.
48. *Designing and Producing Plastic Gears -Plastic Gear Design and Development Procedure*, Celanese, Florence, KY, 2013, pages 10-12.
49. Z. Smith and A. Ulrich, *Exploiting the Potential of Plastics Gears*, Ticona Engineering Polymers, Florence, KY, 2012,Page 12.
50. *General Design Principles for Dupont Engineering Polymers -Design Guide Module 1* E.I. du Pont de Nemours and Company,Wilmington, DE, 1988, page 61.
51. *Vespel*$^R$ Design and Technical Data,E.I. du Pont de Nemours and Company, Wilmington, DE, 2000, page 4.
52. *Designing With Celcon*, Celanese Ticona, 90 Morris Avenue, Summit NJ, page 70.
53. B. Gerjets, *presentation on Fundamentals of Thermoplastic Wear and Friction,* RTP Company. Winona, MN, page 36.
54. ASTM D 2837-01a, *Standard Test Method for Obtaining Hydrostatic Design Basis For Thermoplastic Pipe Materials.*
55. *TR-4 HDB/HDS/SDB/PDB/MRS Listed Materials,* Plastics Pipe Institute$^R$,105 Decker Court, Suite 825, Irving, TX 75062, 9/3/2015.
56. ISO 12162:1995, *Thermoplastics Materials for Pipes and Fittings for Pressure Application — Classification and Designation — Overall Service (Design) Coefficient.*
57. *Planning Fundamentals for Industrial Piping Systems*, Georg Fischer Piping Systems, Shaffhausen, Switzerland, 2006, page 168.
58. *Thermoplastic Industrial Piping Systems Workbook*, Plastics Pipe and Fitting Association, 800 Roosevelt Road, Glen Ellyn, IL, 2005, page 27.
59. ISO 13760:1998,*Plastic Pipes for the Conveyance of Fluids Under Pressure - Miner's rule - Calculation method for cumulative damage.*
60. *Planning Fundamentals for Industrial Piping Systems*, Georg Fischer Piping Systems, Shaffhausen, Switzerland, 2006, page 179.
61. *Designing with Celcon*, Ticona Celanese, 90 Morris Avenue, Summit, NJ, 2002, page 17.
62. P. Tuschak, Pressure Vessel Design for Engineering Plastics, *Engineering Design*, Winter 1984, page 2.

63. *General Design Principles for Dupont Engineering Polymers -Design Guide Module 1* E.I. du Pont de Nemours and Company, 2000, page 20.
64. J.A. Richards, F.W. Sears, M.R. Wehr, and M.W. Zemansky, *Modern University Physics*, pages 630-635, Addison Wesley Publishing Company 1960.
65. D.V. Rosato, D.P. Dimattia, D.V. Rosato, *Designing with Plastics and Composites*, Van Nostrand Reinhold, p 358.
66. *Design Solutions Guide*, BASF Corporation, Florham Park, New Jersey, 2007, page V-5.
67. *Innovative Plastics Vibrational Welding*, Saudi Basic Industries Corporation (SABIC), Pittsfield, MA. 2014, page 14.
68. *Thermoplastic Industrial Piping Systems Workbook*, Plastics Pipe and Fitting Association, 800 Roosevelt Road, Glen Ellyn, IL, 2005, page 53.

# Index

Acetal, 46, 47, 179, 180, 188, 189, 194, 195, 207, 220, 226, 240–242
Acrylic. See polymethyl methacrylate
Acrylic styrene acrylonitrile, 38
Acrylonitrile Butadiene Styrene, 38, 64
Addition polymerization, 5, 6, 8, 32, 35, 37, 39, 42, 60
Additive manufacturing, 90, 92
Adhesives, 275
Aliphatic, 12, 54
Amorphous, 16, 18, 20, 21, 35, 37, 39, 43, 48, 50, 53, 87, 159, 247, 286
Angle of incidence, 247, 249, 254
Angle of reflection, 248
Anisotropic, 93, 231, 243
Annular snap fit, 171
Apparent modulus, 122, 126, 128, 129
Aramid fibers, 88, 232
Aromatic, 12, 37, 51, 52, 54, 111
Arrhenius equation, 159
Average molecular weight, 13–15

Backlash, 202
Barrier property, 159
Barth Velocity Factor, 207
Bearings, 48, 52, 53, 63, 214, 223, 225, 228–231
Biobased polymers, 55
Biodegradeable, 55
Biopolymers, 55
Blister pack, 163
Butt fusion, 272, 273

CAFE. See corporate average fuel economy
Cantilever beam, 127, 170, 172, 175, 182, 197, 203
Carbon fibers, 64, 232
Casting, 92
Cellulose Acetate, 42
Cellulose Acetate Butyrate, 42
Charpy test, 27, 136
Chemical blowing agent, 61, 75
Chemical resistance, 12, 35, 38–40, 49, 50, 53, 61, 62, 93, 110, 223
Chlorinated polyvinyl chloride, 40
CLTE, 18, 19
Coefficient of friction, 46, 47, 49, 53, 63, 169, 180, 181, 183–185, 188, 197, 221, 222, 224
Coefficient of linear thermal expansion. See CLTE
Coefficients of friction, 196, 223, 224, 230
Co-extrusion, 69
Coining, 192, 194, 256
Co-injection Molding, 74
Compression molding, 81, 82
Condensation polymerization, 6, 45, 57, 60, 61
Corporate Average Fuel Economy, 4, 64

Covalent bonding, 12
Creep, 15, 23, 95, 119–122, 124, 125, 127–129, 223–235, 242
Crosslinking, 5, 34, 56
Crystalline, 16, 18, 20, 21, 36, 49, 50, 52, 87, 94, 95, 137, 159, 162
Cyanoacrylates, 276
Cyclic olefin copolymer, 35

Dart Drop Impact Test, 158
Deflection Temperature Under Load. See HDT
Degradation temperature, 23
Design process, 53, 94, 109, 112, 127, 137, 182, 240, 281, 283, 286, 288, 291
Diametrical pitch, 203–205, 212
Dielectric Strength, 28
Draft, 98, 99, 165
Draft angles, 287
Drop weight test, 136

Elmendorf tear test, 158
EMI shielding, 152, 154, 286
Enclosures, 139, 158
Energy director, 269, 270
Engineering plastics, 31, 41
Environmental stress cracking resistance, 110
Epoxies, 59, 60
Ethylene vinyl acetate, 12, 162
Extrusion, 34, 41, 49, 54, 66, 67, 68, 69, 74, 76–79, 90, 109, 162, 164, 166, 192, 194, 244, 256
Extrusion blow molding, 76

FDM. See fused deposition modeling
Filament winding, 87, 89, 244
Finite element analysis, 64, 186, 239, 240, 243, 281
flammability, 8, 29, 37, 39, 40, 142, 149, 152, 284
Flash trap, 270, 275
Flash traps, 272
Fresnell equation, 249
Fused deposition modeling, 90

Gas assist injection molding, 145
Gas Assisted Injection Molding, 75
Gas Counterpressure Foam Molding, 76
Gear Mesh Conditions, 213
Gears, 18, 34, 36, 46, 47, 48, 52, 53, 63, 195, 196, 197, 198, 199, 200–204, 208, 210–215, 217, 218, 221, 222, 286
Glass transition temperature, 20

HDT, 23, 26, 38, 40, 50, 51, 53, 56
Heat stabilizers, 63
Hertz equation, 209
Hoop stress, 234, 237, 240, 242
Hot melt adhesives, 34, 277
Hot plate welding, 273
Hydroscopic, 43, 46, 95, 110
Hydrostatic design basis, 234

Impact strength, 28, 33, 37, 40–44, 47, 48, 49, 51, 62, 134, 240, 285
Index of refraction, 248, 249, 250, 254, 257
Infrared radiant butt fusion, 273
Injection blow molding, 78
Injection molding, 69, 193
Isochronous stress-strain curve, 122
Isotropic, 64, 93
Izod Impact, 135

John Wesley Hyatt, 3, 41

Laser printing, 166
LCP, 52, 53
Lenses, 35, 43, 44, 108, 209, 247, 252, 257
Leo Baekeland, 4, 57
Lewis equation, 203, 208
Light pipes, 247
Light transmission, 35, 43, 107, 247

Limiting Oxygen Index, 29
Liquid crystal polymers. See LCP
Living hinge, 192

Melamine Formaldehyde, 58, 59
Melt Flow Index, 16
Metallocene catalysts, 4, 36
Miner's Rule, 237
Mirrors, 108, 247, 252
Molded in inserts, 267
Molded threads. See threads
Molecular weight distribution, 13, 67
Moment of inertia, 123, 127, 131, 143, 144, 146, 174, 295
Monomer, 5, 6, 8, 11–14, 37, 43, 45, 57

Nylon, 4, 45, 46, 51, 64, 77, 90, 95, 110, 159, 193, 194, 195, 220–222, 226, 228, 231, 242

Oil canning, 144
Oxidation, 111

Parison, 76, 77, 162, 164
PBS. polybutylene succinate
PC-ABS
Polycarbonate/ABS, 44
PEEK, 51, 54, 220
Permeability, 12, 48, 159, 162–164
PHA, 56
Phenolic, 4, 58
Phenolics, 57, 222
Pinion gear, 197
Pipe, 9, 18, 33, 38–40, 46, 49, 51, 60, 61, 63, 67, 69, 88, 110, 117, 234, 235, 237–239, 241, 245, 249, 254
Pipe threads, 263
Pitch circle, 203
Pitch line, 197, 201, 203, 204, 207, 208
Pitch Point, 203
PLA, 55, 56, 90
Poisson's ratio, 133, 194, 209
Polar bonding, 12
Polar moment of inertia, 132, 133
Polybenzimidazole, 53
Polybutylene succinate, 56
Polybutylene terephthalate, 47
Polycarbonate, 32, 44, 121, 125, 176, 182, 183, 254
Polydicyclopentadiene, 62
Polyether ether ketone. See PEEK
Polyethylene, 9, 13, 17, 32–35, 39, 47, 48, 61, 84, 110, 111, 159, 165, 193, 222, 230
Polyethylene, 32, 35
Polymethyl methacrylate, 42, 111
Polyoxymethylene. See acetal
Polyphenylene sulfide. See PPS
Polypropylene, 35, 37, 193
Polystyrene, 37, 38
Polysulfone, 50
Polytetrafluoroethylene, 4, 17, 48, 230
Polytetrafluoroethylene, 222
Polyurethanes, 61, 85, 222, 276
Polyvinylidene chloride, 40
PPS, 50
Press fits, 265
Press motion, 190, 191
Pressure Angle, 203
Product life, 284
Proportionality constant, 176, 179, 183
Prototyping, 60, 86, 90, 92, 165, 166, 207, 217, 281, 288
Pultrusion, 89
PVC. See polyvinyl chloride

Reaction injection molding, 85
Recycling, 112
Reflection, 247, 252, 254
Refractive index, 248, 249, 254
Relative thermal index, 138
RIM. See Reaction Injection Molding
Rotational molding, 82
Roy Plunkett, 4

SAN. See styrene acrylonitrile
Secant modulus, 119, 181–184, 188
Selective Laser Sintering, 91
Self-tapping screws, 263
Shear joint, 269, 270
Shrinkage, 50, 52, 63, 73, 75, 93, 94, 95, 96, 99, 101, 137, 140, 145, 209, 216, 218–220, 228, 229, 231, 243, 256, 287
Silicones, 276
Skin pack, 163
SLA, 90, 91
S-N curve, 130
Snap fits, 169, 170
Snell's law, 249, 254
Solid phase pressure forming, 87
Solvent bonding, 275, 277
Spin welding, 274
Steric effects, 17
Steric hindrance, 35
Stretch blow molding, 79
Styrene acrylonitrile, 38
Surface resistivity, 155
Surface veil, 89

Talc, 63
Thermal conductivity, 7, 19, 20, 37, 146
Thermoforming, 85
Thermoplastic Elastomers, 54
Thermoplastic polyimides, 53
Thermoplastics, 4, 5, 23, 31, 32, 41, 52, 57, 67, 73, 77, 81, 82, 84, 87, 93, 101, 169, 174, 192, 224
Thermoset polyurethanes, 61
Thermosets, 5, 6, 24, 45, 58, 69, 76, 84, 85, 87, 91, 96, 179, 199
Threads, 79, 84, 260, 261–264
Tip relief, 212
Tolerances, 87, 93–96, 207, 209, 211, 214, 215, 221, 228, 256, 257, 287
Tongue and groove joint, 275
Torsion, 131
Total composite error, 210
Transfer molding, 82
Tribology, 223

UL Yellow Card, 152
Ultrasonic welding, 267, 268, 272
Ultrasonic welding, 268
Unsaturated polyesters, 60
Urea formaldehyde, 4
Urea Formaldehyde, 58
UV stabilizers, 286

Vibration welding, 272
Viscoelastic behavior, 87, 117, 123, 196
Viscosity, 5, 14, 52, 75
Voice of the customer, 282
Volume resistivity, 153, 155

Wallace Carothers, 4
Wear, 225, 300
Weathering, 93, 111, 117, 138, 233, 286
Weathering, 111, 138
Weld lines, 73, 109, 131, 136, 140, 194, 220, 230
Worm gears, 200

Yield point, 25, 118, 175
Young's modulus, 25, 119, 182

Zeigler Natta catalysts, 4